DATE DUE

Basics of
Electrical
Power Transmission

Anthony J. Pansini, E.E., P.E.

Life Fellow IEEE; Member ASTM

Engineering and Management
Consultant
Waco, Texas

Prentice Hall, Englewood Cliffs, New Jersey 07632

9-93

19971825

Library of Congress Cataloging-in-Publication Data

PANSINI, ANTHONY J.
    Basics of electrical power transmission / Anthony J. Pansini.
        p.   cm.
    Includes index.
    ISBN 0-13-059866-6
    1. Electric power transmission. I. Title.
TK3001.P27 1990
621.319--dc20

89-16043
CIP

Editorial/Production supervision
  and interior design: *Jean Lapidus*
Cover design: *Ben Santora*
Manufacturing buyer: *Ray Sintel*

 © 1990 by Prentice-Hall, Inc.
A Division of Simon & Schuster
Englewood Cliffs, New Jersey 07632

This book can be made available to businesses
and organizations at a special discount when
ordered in large quantities. For more information
contact:

Prentice-Hall, Inc.
Special Sales and Markets
College Division
Englewood Cliffs, N.J. 07632

Printed in the United States of America

10 9 8 7 6 5 4 3 2 1

ISBN 0-13-059866-6

PRENTICE-HALL INTERNATIONAL (UK) LIMITED, *LONDON*
PRENTICE-HALL OF AUSTRALIA PTY. LIMITED, *SYDNEY*
PRENTICE-HALL CANADA INC., *TORONTO*
PRENTICE-HALL HISPANOAMERICANA, S.A. *MEXICO*
PRENTICE-HALL OF INDIA PRIVATE LIMITED, *NEW DELHI*
PRENTICE-HALL OF JAPAN, INC., *TOKYO*
SIMON & SCHUSTER ASIA PTE. LTD., *SINGAPORE*
EDITORA PRENTICE-HALL DO BRASIL, LTDA., *RIO DE JANEIRO*

To Harold M. (*Jolly*) Jalonack

Friend and Mentor

Contents

Preface

This book comprises the first revision of the original work of the same title. The revision includes the updating of the material and the expansion of additional material, especially on Extra High Voltage and Direct Current Transmission. Another addition includes a chapter on Basic Electricity that should aid those who may lack some understanding of the characteristics and phenomena associated with electric circuits.

Three appendices also have been added. The first appendix concerns environmental considerations that are assuming ever-greater importance. The second appendix includes a step-by-step guide for uprating transmission lines for higher operating voltages, and can be used for designing structures to operate at high voltages initially. The third appendix contains factors for converting data into metric units, in line with the trend in this country to join the metric fraternity.

Basic information is provided for those just entering this field. Others, more experienced, may discover some valuable nuggets as well as find it a desirable review. This text is also meant to serve the needs of those whose work is associated with this part of the electric industry. It is a training resource for the several categories of employees in the electric utilities, both private and public. It is also a source of much-needed information for people in associated manufacturing and service industries, in schools and educational centers, for planning and civic groups interested in the esthetic and economic development of reliable electric supply, in the financial community of investors and promotors and promoters, for members of the legal profession involved in contracts and litigation of several kinds, and for arbitrators and mediators.

Acknowledgment is made to the many manufacturers and agencies who contributed drawings, photographs, and descriptive material. And to the many who, through the years, have enabled this writer to assemble the information being passed on in this work. To the staff of Prentice Hall for their editing and art work assistance, and especially to Mrs. Jean Hunter Lapidus for her constructive criticisms and suggestions, and for her resourcefulness in meeting unusual problems that arose in preparing this manuscript for publication on schedule. And last, but certainly not least, to my beloved wife for her patience, understanding, and encouragement during the preparation of this work.

Waco, Texas   *Anthony J. Pansini*

1

General
Concepts

The system of supplying electricity to a community may be compared to a tree, each leaf representing a customer or group of customers (Fig. 1.1). The roots of the tree represent the generating equipment, supplying nourishment to all parts of the tree. The trunk of the tree, which carries all the sap or life of the tree, is similar to the transmission line conducting electricity from the generating station to the various substations in different parts of the system. The branches represent the distributing system conducting electricity to each small branch and leaf.

Similar to almost every other business, means must be provided to bring the product or service to the consumer; in this case electricity from the generating plant to the consumer. In most instances, separate provision is made for the delivery of the product in bulk and for its retail distribution. In an electrical power system, the former is referred to as *transmission* while the latter is known as *distribution*. (See Figure 1.2.)

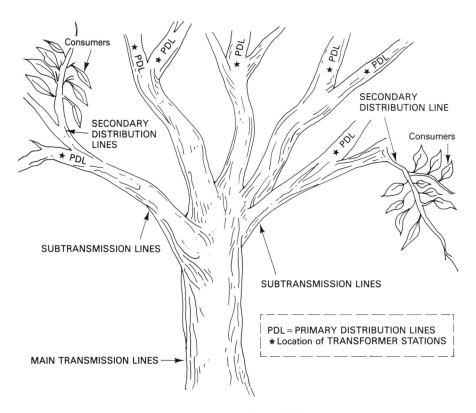

Figure 1.1 Tree of electricity.

THE TRANSMISSION SYSTEM

Transmission systems serve to carry the bulk supply to load centers from generating plants generally located in outlying areas because of their environmental impact and need for large amounts of water for cooling purposes. A factory or generating plant is rarely situated at or near a population or load center. Generally, high voltages and great capacity are essential with the transmission system.

Transmission lines also provide interconnections for the transfer of electrical power between two or more utilities for economic and emergency purposes. Such interconnections of utilities are referred to as pools or grids, and sometimes more formally as integrated systems.

From wholesale centers or transmission substations, means are provided to distribute the product, electricity, to regional load and population centers where a demand for the product exists. These means constitute the distribution system.

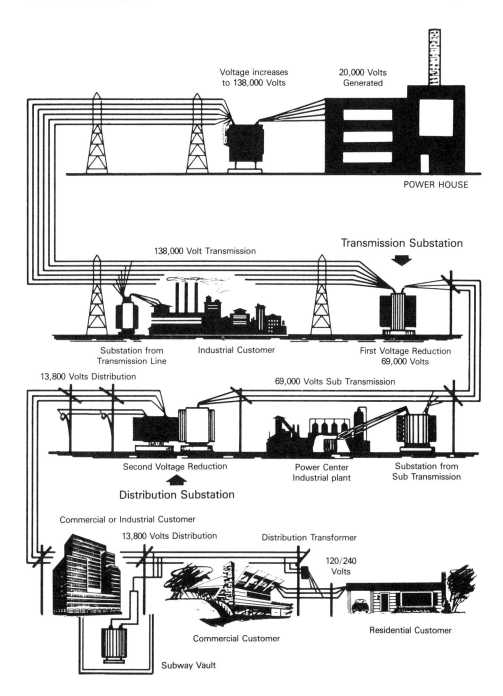

Voltage increases
to 138,000 Volts

20,000 Volts
Generated

POWER HOUSE

Transmission Substation

138,000 Volt Transmission

Substation from
Transmission Line

Industrial Customer

First Voltage Reduction
69,000 Volts

13,800 Volts Distribution

69,000 Volts Sub Transmission

Second Voltage Reduction

Power Center
Industrial plant

Substation from
Sub Transmission

Distribution Substation

Commercial or Industrial Customer

13,800 Volts Distribution

Distribution Transformer

120/240
Volts

Residential Customer

Commercial Customer

Subway Vault

Figure 1.2 The electrical supply system: electrical service from the
generator to the customer.

The principal purpose of a transmission system, then, is to carry bulk quantities of electrical energy to or between convenient points. At these points, the electrical energy may be subdivided for eventual delivery to one or more distribution systems. A synthesis of definitions accepted by the Federal Power Commission and various State Commissions follows:

> A transmission system includes all land, conversion structures and equipment at a primary source of supply; lines, switching and conversion stations between a generating or receiving point and the entrance to a distribution center or wholesale point; all lines and equipment whose primary purpose is to augment, integrate or tie together sources of power supply.

The Federal Power Commission also emphasizes the significance of transmission systems in a survey.

> The strategic importance of transmission is much greater than is indicated by the twenty percent average share in the overall cost of electricity. Adequate interconnections, where economically justified, provide the keys to large scale generating units, to major savings in capacity because of load diversity, and the most efficient utilization of existing generating capacity. In short, interconnection is the coordinating medium that makes possible the most efficient use of facilities in any areas or region.

In the pictorial rendition, note that the generator produces electricity at a pressure of 20,000 volts. This pressure is raised, by means of a piece of apparatus known as a transformer, to a value of 138,000 volts or higher for the long transmission journey. This electric power is conducted over 138,000-volt (or 138-kV) transmission lines to substations located in important centers of population or electrical loads in the territory served. When the electric power reaches the substations, its pressure is reduced or stepped down (also by a transformer) to 69,000 volts (or 69kV) for transmission in smaller quantities to other substations in the local load areas. (In some cases, it might be stepped down to 13,800 volts [13.8kV] for direct distribution to local areas.) Transmission circuits of such voltages may consist of open wires on poles in outlying zones (along highways, for example) where this type of construction is practical, or may consist of cables installed in ducts underground or buried directly in the ground in more densely populated areas.

While the transmission line system shown in Figure 1.2 is simple, many transmission line installations can provide for an interchange of power between two or more neighboring utility companies to their mutual advantage. Such lines permit one utility to purchase power from its neighbors when it is economically advantageous in preference to its own generating facilities, or during periods of emergency when its own generating units may be out of service for repair or maintenance. Such interconnections between utility companies may be very complex, and are often referred to as a grid or pool (Fig. 1.3).

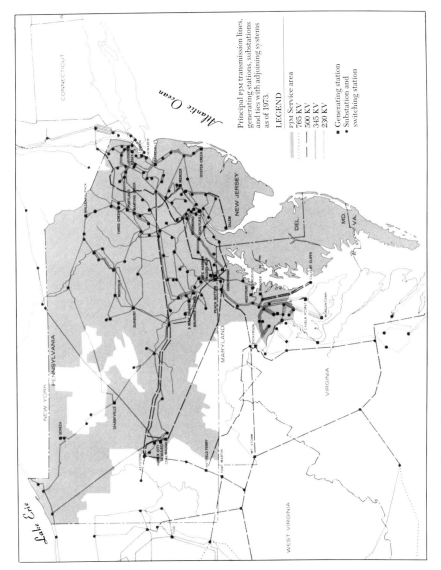

Figure 1.3 (a) Major transmission grid in the northeast United States.
(b) Major transmission lines in the United States.

5

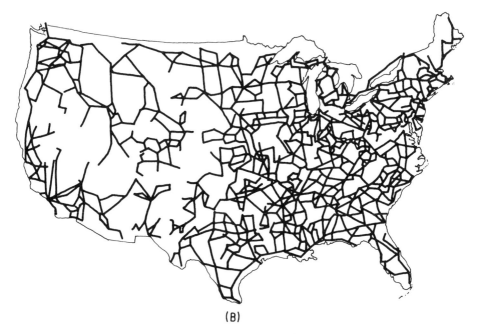

(B)

Figure 1.3(b) *(continued)*

Wheeling of Power

While the transmission interconnections are obviously made between contiguous neighboring utilities, the transfer of power is not limited to utilities adjacent to each other. Power may be interchanged between utilities remote from each other by using the transmission facilities of the intervening utilities; this is referred to as "wheeling" (sometimes "wheel-barrowing") of the power, with appropriate compensation provided for the use of their facilities to all the utilities involved and usually included in the contractual arrangements establishing membership in the grid or pool.

LINE CHARACTERISTICS

Before discussing the physical and electrical characteristics of transmission lines, it may be well to consider the analogy with water systems.

Water Current Analogy

The flow of electricity can be compared to the flow of water. Where water flows in a pipe, electric current is made to flow through conductors or wires. (See Figure 1.4.)

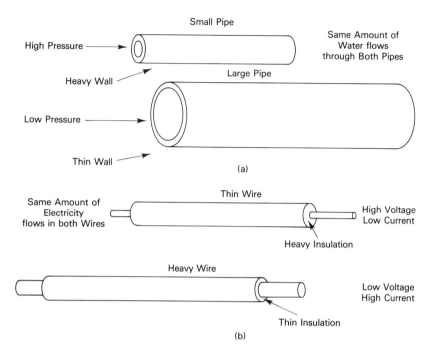

Figure 1.4 Water—current flow analogy. (a) Comparison of water flow through different size pipes. (b) Comparison of current flow in different size wires.

To move a definite amount of water from point to point in a given amount of time, a large-diameter pipe and a low pressure may be employed to force it through, or a small-diameter pipe and high pressure applied to the water. If the higher pressure is used, the pipe must have thicker walls to withstand the pressure.

The same rules apply to the transmission of electric current. In this case, the diameter of the pipe corresponds to the diameter of the wire, and the thickness of the pipe walls corresponds to the thickness of the insulation around the wire. (For more detailed information, refer to Chapter 6, Basic Electricity.)

Voltage and Load Handling Ability

Transmission lines are designed to transport relatively large amounts of electric power, usually expressed in watts, kilowatts (1000 watts), or megawatts (millions of watts), over relatively long distances. These may range from less than 10,000 kilowatts and a few miles to over several million kilowatts and many hundred miles. The electrical pressures, or voltages, at which these lines operate may range from a few thousand volts to values of over 765,000 volts (or 765 kV), with experimental lines operating at 1500 kV. In general, the larger the amount of power to be carried, and the greater the distance to be traversed, the higher the voltage at which the transmission line is designed to operate (Fig. 1.5).

Historically, standard transmission voltages have been approximate multiplies of a base of 115 volts.

2,300	34,500	354,000
4,600	46,000	500,000
6,900	69,000	765,000
11,500	115,000	1,000,000
13,800	138,000	1,500,000
23,000	230,000	

Transmission lines between substations and local or regional load centers are sometimes called subtransmission lines. They generally operate at voltages of 138,000 volts or less. Lines operating at 500,000 volts and above are generally referred to as Extra-High-Voltage (EHV) lines.

Cost Considerations

Transmission facilities are costly, whether overhead or underground. The investment required, as a percentage of total utility plant, may be as much as 20 percent. This is not only because of the requirement to move even larger amounts of power over considerable distances, but also because of the greater attention being given to appearance and effect on environment. Transmission

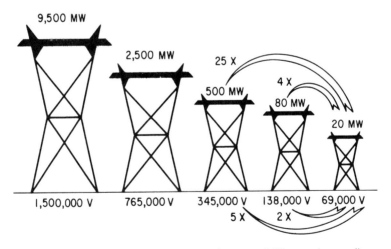

Figure 1.5 Comparison of power-carrying capabilities and operating voltages.

structures have been specifically designed for appearance, employing gracefully shaped and colored structures pleasing to the eye and blending with the surroundings (Fig. 1.6). The color of insulators has generally been brown, but changes to light blue and gray standards have been made. These colors are less obtrusive against sky backgrounds and esthetically more acceptable.

Rights-of-Way

Transmission costs are affected by the costs of rights-of-way, the higher quality of construction in congested areas, and the need for stronger ties both within and between utilities because of reliability requirements introduced by the use of larger generating units and underground cable systems. Transmission requirements may be reduced, however, when generating plants are in or near the load centers. Data collected by the Federal Power Commission indicates costs for underground transmission lines in rural and suburban areas may vary from 10 to 30 times as much as overhead lines, depending on the types of construction (duct or direct burial) and cables (solid insulation or hollow type) employed. In urban areas, the ratio may be greater. In other places, because of rough ground, inaccessible terrain, and special situations, this cost may be as much as 50 times greater. (See Figure 1.7.) It would appear, therefore, that overhead lines may continue to constitute a predominant part of transmission systems for some time to come.

Rights-of-Way and Appearance

Overhead transmission lines are admirably suited to open country areas where rights-of-way of ample width and reasonably straight lines are available (Fig. 1.8). The straight line route, with few angles, decreases the length of line required and the amount of guying or special construction necessary. Further, these rights-of-way are usually easier to acquire, whether by purchase, lease, or other arrangement. Appearance may not always be of paramount importance. The heights of transmission lines required by national and local safety code clearances for the voltage ranges involved and the economics of long spans make such overhead installations very adaptable to open cross-country areas.

If transmission lines must be built in densely populated urban and suburban areas, rights-of-way may be restricted to relatively narrow areas along the streets or in alleys in the rear of buildings. Construction is usually limited to single, rather high, poles with span lengths of one city block or less. Unusual attention must be given to appearance, tree conditions, guying, and other factors which may be detrimental to good public relations. This may also include limited access in which materials may have to be transported by hand as well as agreements limiting work to off-schedule hours.

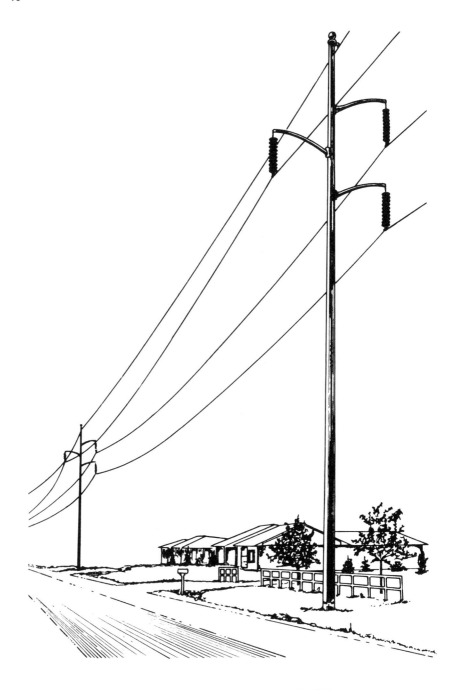

Figure 1.6 Pole structures in residential areas.

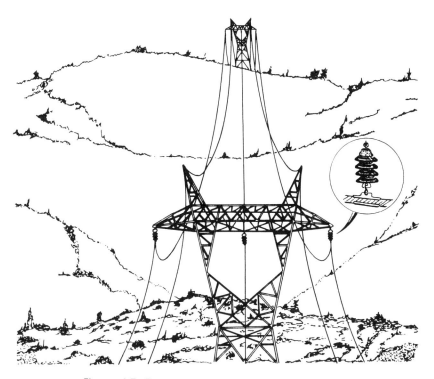

Figure 1.7 Transmission lines in inaccessible areas.

Tree Clearing

Often the rights-of-way require clearing of trees, brush, and other growth, and must be maintained sufficiently clear so that subsequent growth will not affect operation of the lines. The engineer will designate all danger trees which may be removed or topped at the option of the contractor. In approximately level terrain, trees which would reach within 5 feet of a point underneath the outside conductor in falling are examples of danger timber. In wooded areas, the right-of-way is cleared back far enough so that trees falling will not fall into the line; trees may be topped gradually so that so-called danger timber will not inflict damage to the lines (Fig. 1.9). Access roads must also be provided not only for construction, but for future patrols, inspections, repairs, and maintenance of the lines. Portions of the right-of-way must be cut so that stumps will not prevent the passage of tractors and trucks along the right-of-way.

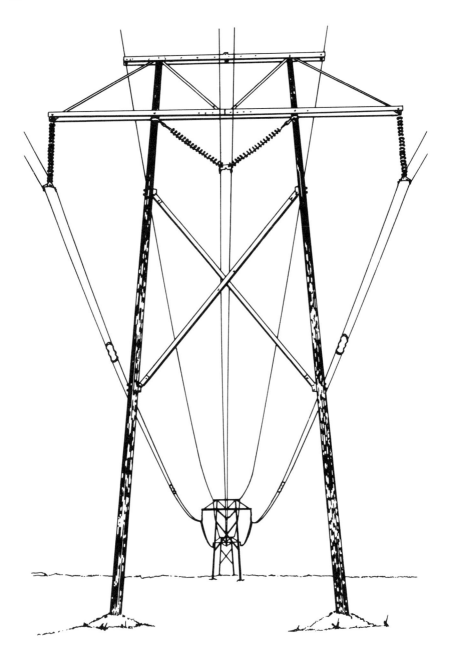

Figure 1.8 Line installation in open, flat terrain.

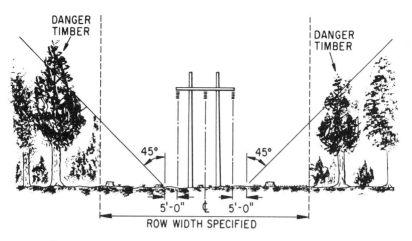

Figure 1.9 Tree removal and topping along right-of-way.

Rough Terrain

In some instances, such as where lines traverse heavily treed rough terrain, helicopters may be used to deliver materials, often preassembled tower structures, as well as personnel to the job sites for both construction and maintenance purposes (Fig. 1.10).

Railroad Rights-of-Way

The installation of transmission lines along railroad rights-of-way would appear to provide a logical and desirable location for such lines (Fig. 1.11). Often, however, these rights-of-way may also contain communication circuits which may suffer electrical interference from the high voltages of the transmission lines. Then, too, soot and smoke from coal or oil fired locomotives contaminate the surfaces of the insulators overhead, subjecting them to flashover, unless periodically cleaned. Care must be exercised during work on such lines to insure the safety of personnel from passing trains. This may sometimes cause protracted and expensive delays in work schedules. Despite these drawbacks, railroad rights-of-way provide excellent routes for transmission lines, especially as the railroad usually serves population or industrial centers which also constitute electric load centers.

Increasing Capacity

Overhead transmission lines lend themselves to growing electric systems. Their capability may be readily increased by raising their operating voltages

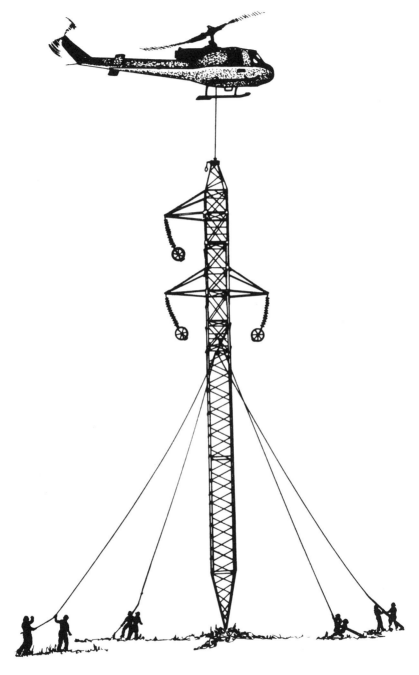

Figure 1.10 Helicopters used to help install towers in inaccessible areas.

Figure 1.11 Transmission line (left) and low-voltage communications line (right) along railroad right-of-way.

usually by only adding or replacing insulators or by adding or replacing conductors. Additional lines or substations can be connected to existing systems quite readily.

Underground transmission lines are generally limited in capacity by the initial installation. Increases in capability are generally accomplished by the addition of other parallel lines or the replacement of the original cable system, both of which are very costly processes.

Overhead Versus Underground Repairs

While overhead lines are exposed to the hazards of vandalism and weather, particularly of lightning, ice, and wind, faults are usually relatively easy to find and repair. The reverse is generally true of underground lines. While relatively free from harmful exposure, when faults do occur, they are relatively hard (and time consuming) to find and repair. Highly specialized skills are required for both their installation and repair. The addition of new lines or substations in an underground transmission system presents difficulties not usually met in overhead systems.

DETERMINING TRANSMISSION VOLTAGES

Earlier, the relationship between conductor sizes and voltages in the transmission of electrical power was explained, including an analogy with water carrying systems. This principle is basic in considering the choice of a voltage (or pressure) for a transmission system.

There are two general ways of transmitting electricity: overhead and underground. In both cases, copper or aluminum conductors, or aluminum conductors steel reinforced (ACSR), and, in some cases, steel conductors, are used. But the insulation in the first instance is usually air except at the supports (towers, poles, or other structures), where it may be porcelain, glass, or other material. In underground transmission, the conductor is usually insulated with oil impregnated paper, or a special type of plastic material.

In overhead construction, the cost of the copper or aluminum conductors as compared to the insulation is relatively high. It is therefore desirable, when transmitting large amounts of electric power, to resort to the higher electrical pressures or voltages, thereby employing thinner, less expensive conductors that are easier to handle. Low voltages require heavy conductors that are costly and bulky and expensive to install.

There is a limit, however, to how high the voltage and how thin the conductors can be. In overhead construction there is the problem of supports — poles, structures, towers. If the conductor is made too thin, it will not be able to support itself mechanically and the cost of additional supports and insulators becomes inordinately high. Underground construction faces the same economic limitations, and in this case, the expense of insulation. Underground, a cable must be thoroughly insulated and sheathed from corrosion. The greater the overall size of the cable, the more sheathing becomes necessary and more difficulty experienced in its handling (Fig. 1.12).

Determining transmission voltages is a matter that requires careful study. Engineers cost out the systems employing several generally standard voltages (and standard materials and equipment). For example, the annual costs for each of the systems: 69 kV (69,000 volt), 138 kV, and 230 kV. Approximate costs of conductors and necessary equipment, insulators, supports, rights-of-way, etc., including the labor, and associated expenses involved in their construction and installation, are carefully evaluated. The annual carrying charges are calculated, to which are added the estimated annual maintenance

Overhead

Thick conductors—lower voltage—longer spans—fewer supports and insulators.

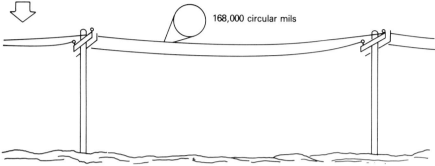

168,000 circular mils

Thin conductors—higher voltage—shorter spans—more supports and insulators.

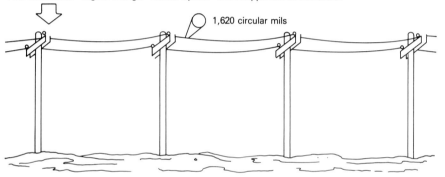

1,620 circular mils

High Voltages require more insulation—more sheathing
—lower voltages thus may prove economical

Underground

conductor   insulation   sheath

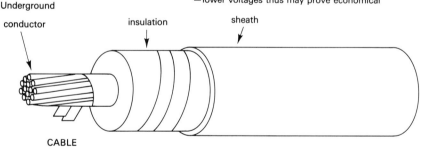

CABLE

Figure 1.12 Practical economics affect the size of a transmission line.

expenses and the value of the losses for each of the systems under study; the lowest sum of these annual costs usually determines the selection of the voltage. The lowest overall annual expense will generally be found to occur when the annual carrying charges are approximately equal to the annual cost of the electrical losses incurred; this is known as Kelvin's Law.

In making these determinations, the future with its possibilities of increased demand is also taken into consideration. The increase over a period of time is estimated and is included in the study. Again, economic studies determine the extent of the future time period to be considered, at the end of which facilities would need to be reinforced or replaced.

In the calculations involved in arriving at the selection of the transmission voltage, the construction, maintenance, and operation of the projected facilities are subject to the overriding considerations of safety. In general, the utility standards are higher than those recommended by the National Electric Safety Code and those imposed by local and governmental regulations.

REVIEW

- [ ] The system of supplying electricity to a community includes facilities for production (generating plants), for wholesale or bulk delivery (transmission), and for local retailing (distribution).

- [ ] Transmission circuits may consist of open wires on towers or poles in outlying zones, or of cables installed in ducts underground or buried directly in the ground in more densely populated areas. Transmission line installations can provide for an interchange of power between two or more neighboring utility companies to their mutual advantage. Power may also be "wheeled" to non-adjacent companies over the facilities of the intervening companies. Such interconnections between utility companies are often referred to as a grid or pool (Fig. 1.3).

- [ ] Transmission lines transport relatively large amounts of electric power over relatively long distances. Power may range from less than 10,000 kilowatts and a few miles to over several million kilowatts and many hundred miles. The voltages at which these lines operate range from a few thousand volts to over 765,000 volts.

- [ ] The voltage of a transmission line is usually determined by economics. Higher voltages are generally desirable as smaller, less expensive, conductors may be used; however, for overhead lines, the mechanical strength of the conductor may limit the span length, requiring a greater number of poles, insulators, etc. The annual costs of several standard voltage lines are evaluated and that having the lowest annual cost usually determines the selection of voltage.

- [ ] Transmission lines between substations and local or regional load centers are sometimes called subtransmission lines. They generally operate at voltages of 138,000 volts or less.

- [ ] Lines operating at 500,000 volts and above are generally referred to as Extra-High-Voltage (EHV) lines.

☐ Overhead transmission lines are more suited to open country areas where rights-of-way of ample width and reasonably straight lines are available. Appearance also is not always of paramount importance (Fig. 1.8).

☐ While overhead lines are exposed to the hazards of vandalism and weather, faults are usually relatively easy to find and repair. Conversely, underground lines, while relatively free from harmful exposure, are relatively hard (and time-consuming) to find and repair, when faults do occur.

STUDY QUESTIONS

1. What is the function of transmission lines or systems?
2. What factors influence the voltage selected for a transmission line?
3. How does the load carrying capability of a transmission line relate to the voltage at which the line operates?
4. What are some typical standard voltages of transmission lines?
5. What is meant by subtransmission? At what voltages do they operate?
6. How do generating plants affect the transmission system?
7. What are the advantages of overhead constructed transmission lines?
8. What are the disadvantages of overhead constructed transmission lines?
9. What are the advantages of underground constructed transmission lines?
10. What are the disadvantages of underground constructed transmission lines?

2

Overhead Construction

The several elements constituting an overhead transmission line, together with pertinent information concerning their purpose, characteristics, and methods of handling, are covered in this chapter.

Poles and Towers

Structures for supporting the overhead conductors are broadly classified as poles or towers [Fig. 2.1(a),(b),(c),(d)].

Single pole supports may be spaced a few or more hundred feet apart. They may be made of natural wood: southern yellow pine, western red cedar, douglas fir, larch, and other species, reflecting the availability of suitable timber in the several geographic areas of the country, or of imported Wallaba where extra strength is required [Fig. 2.1(a)]. These may be chemically treated round shaped tapered poles, or square shaped poles. Poles may also be of hollow tapered tubular design, made of steel or aluminum; they may also be built up of flat metal members, latticed together, into a variety of cross sections, from squares to duodecagons (12-sided) [Fig. 2.1(b),(d)]. Wood poles are set directly in the ground. Metal poles may be embedded directly in earth or concrete, or secured to bolts embedded in a concrete base. Wood and metal poles may be combined as members of a structure into A-frames, H-frames,

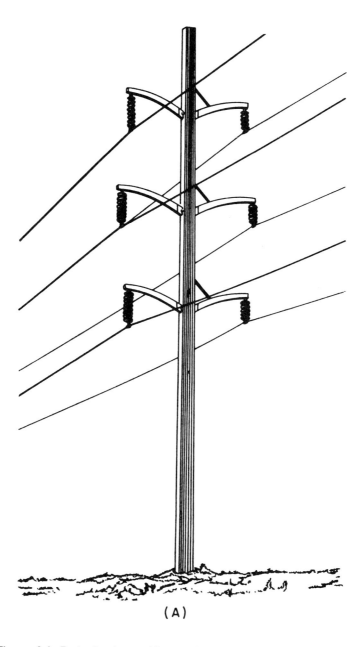

(A)

Figure 2.1 Typical poles and towers. (a) Wood pole; (b) Hollow, tapered, tubular pole; (c) Prestressed concrete transmission tower; (d) Steel lattice tower.

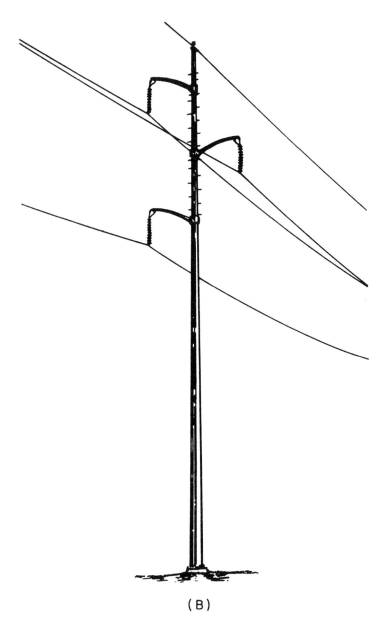

(B)

Figure 2.1(b) *(continued)*

and sometimes into V- or Y-type transmission line supports capable of carry-ing longer spans [Fig. 2.1(c)].

Reinforced concrete, hollow, round, or square poles have long life, structural strength, freedom from weathering, good appearance, and low

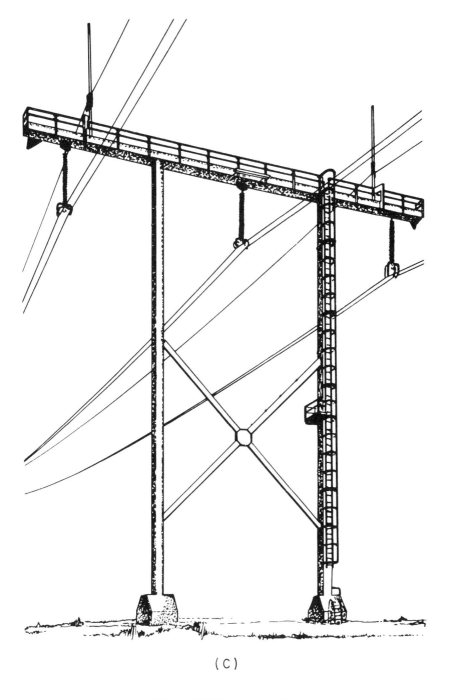

(C)

Figure 2.1(c) *(continued)*

(D)

Figure 2.1(d) *(continued)*

maintenance costs. They are, however, relatively expensive in first cost and extremely heavy, making their handling difficult.

Tower Types

Common long-span-type construction consists of structures suitably latticed or braced, of galvanized steel or aluminum, mounted on wide rectangular bases, and sometimes referred to as towers (Table 2.1). Their footings may be of reinforced concrete or grillages made of assembled steel beams. Towers may be of several types: a tangent tower on which the conductor supported is essentially a straight line; a small-angle tower on which the conductor supported changes direction slightly, perhaps 5° or 10°; a medium-angle tower

TABLE 2.1 Typical Pole Setting Depths

Pole Length (ft)	Approximate Setting Depth	
	All Solid Rock (ft)	Firm Soil (ft)*
35	4.5	6.0
40	5.0	6.0
45	5.5	6.5
50	6.0	7.0
55	6.0	7.5
60	6.5	8.0
65	7.0	8.5
70	7.0	9.0
75	7.0	9.5

*Based on approximately 10 percent of pole length plus 2 feet.

with support changes of 20° to 30°; and a heavy-angle tower which accommo-dates sharper turns. Dead-end towers support the entire pull of conductors which terminate on these towers; these towers are located at anchor points in a line, at certain angles in the line, or at points of take-off from the line. Structures may be of self-supporting latticed (tower) construction or guyed; these latter may be modified "A," "H," "V," "Y," or guyed-mast types, designed to give a certain flexibility to the structure [Fig. 2.1(d)]. Towers or structures are usually referred to by type, voltage, and number of circuits; for example, H-frame, tangent, 138 kV, and double-circuit tower. In many cases, the transmission line will contain two or more types of supporting structures.

Guys and Angles

In general, guys and anchors are installed where lines may terminate (referred to as dead-ends), at angles, or long spans where pole strength may be exceeded, and at points of excessive unbalanced conductor tension. Guying for line protection is installed at various points in the line to limit damage should storms or other violent action cause the line to fail in somewhat of a domino fashion. While practically all wood or metal pole transmission structures have guy and anchor installations associated with them, occasionally, some self-sustaining towers may also require their installation; for example, at 90° turns in the line [Fig. 2.2(a),(b),(c),(d)]. Both guy wire and anchors used in transmis-sion lines are generally heavier than those for distribution lines.

Distribution Underbuild

In some instances, transmission poles may also carry distribution cir-cuits, generally in a lower position on the pole. Extreme care should be taken in the design of such common lines to protect the distribution lines from the

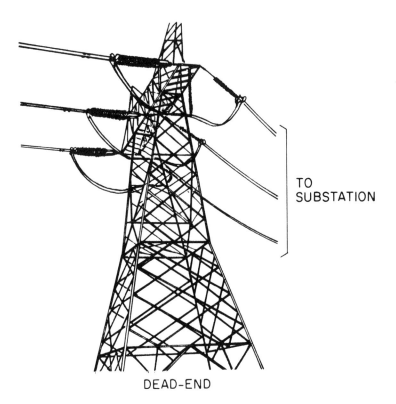

TO
SUBSTATION

DEAD-END

Figure 2.2 (a) Tower constructions; (b) Guyed structures; (c) Wood and metal pole constructions; (d) Tower constructions.

effects of the higher voltage lines and to provide safe working conditions for the workers. Additional guying may be necessary to compensate for the effect of such underbuild on transmission structures (Fig. 2.3).

Design Factors

The choice of supporting structures is influenced by many factors which, considered together, result in the greatest economy. The voltage of the line or lines to be carried, both present and future, determine the spacing of the conductors and the length of the string of insulators required. This may be subject to modification to provide proper space and conditions for the workers. Span lengths in turn are affected by the clearances required for safety (Table 2.2). These minimums are usually specified by the National Electric Safety Code (NESC) whose provisions are established and revised periodically

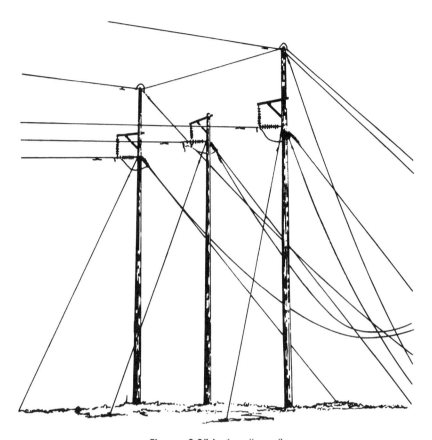

Figure 2.2(b) *(continued)*

by the Institute of Electrical and Electronics Engineers (IEEE) and generally reflects latest practices; reference should therefore be made to the latest code revision. They are also influenced by local ordinances, and by the allowable difference in sags of the conductors under summer and winter conditions. The size, type, and material of the conductor also have an effect. Thus, aluminum conductors generally have a greater summer sag than copper conductors, requiring higher supports to achieve the same minimum clearance above ground and also to keep the conductors from swinging into each other (Table 2.3). Appearance, the character of the terrain (whether plain or hilly), climatic conditions, the nature of the soil, and transportation facilities, all have to be evaluated. The structure itself — wood, metal, or concrete — is also a determining factor.

The first cost of such supports is not the only consideration to be taken

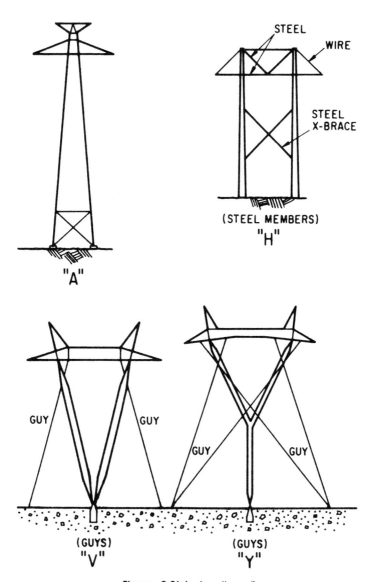

Figure 2.2(c) *(continued)*

into account; the cost of operation and maintenance is also important. The longer the spans, the fewer will be the number of points of support, with fewer insulators necessary requiring possible replacement, fewer structures requiring painting, and fewer poles requiring replacement. As all of these factors may vary with the time of installation, location, local conditions, prevailing regula-

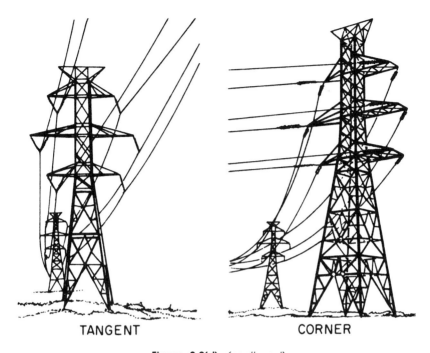

TANGENT   CORNER

Figure 2.2(d) *(continued)*

tions, and with plans for the future (for example, addition of a second line several years after the initial installation), the final decision is usually the result of many judgments and compromises. Computers can make this procedure less toilsome by making many calculations, varying each and every factor as desired. Further, they can do this in the same time that engineers previously calculated by laborious hand methods only a few alternate choices, varying only a few of the factors which, in their judgment, would have the greatest impact on the ultimate solution. In general, the calculated results, either by computer or hand methods, will depend on the correctness of estimates of cost, depreciation allowances, and power to be transmitted.

Cross-arms

Cross-arms for towers are generally of galvanized steel or aluminum, and of wood for pole and A- or H-frame construction. Laminated wood cross-arms and arms made of epoxy synthetics have also been used (Fig. 2.4). With these types, since the material is an insulator, the number of insulators in a string may be reduced, and the other advantages are better appearance and better working conditions because of the completely insulated arm.

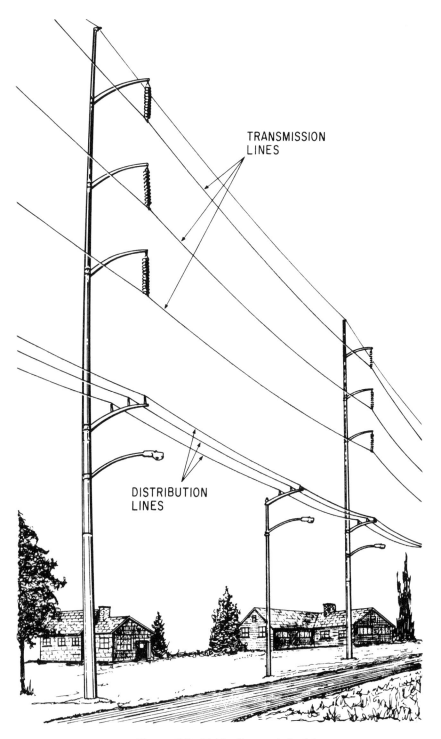

TRANSMISSION
LINES

DISTRIBUTION
LINES

Figure 2.3 Distribution underbuild.

TABLE 2.2 Recommended Minimum Vertical Clearance of Conductors
(in ft) Above Ground or Rails*

(120°F, No Wind, Final Sag. Conductor Design Tension
50% of Ultimate Strength)

Nature of ground or rails underneath	34.5 kV	46 kV	69 kV	115 kV	138 kV	161 kV
Track rails of railroads	32	32	33	35	35	36
Public streets and highways	25	25	26	28	28	29
Areas accessible to pedestrians only	19	19	20	22	23	24
Cultivated fields†	20	20	21	23	24	25
Along roads in rural districts	22	22	23	25	26	27
Natures of wires crossed over						
Communication lines‡	8	8	9	11	11	12
Supply lines up to 50,000 volts‡	6	6	7	9	9	9

*Sag should allow one foot greater clearance than shown above. Conductors smaller than 1/0 ACSR may require additional clearance.

†The NESC does not specify the clearances that should be maintained across cultivated fields. Local conditions and regulations may call for deviations from these recommended minimum clearances.

‡For lower conductor at its initial unload sag.

Always refer to the latest edition of the Code.

Conductors

The size, type, and material of conductors are affected by the amount of power desired to be transported and the distance it is to be transported, both factors influencing largely the voltage at which the line is to operate. Theoretical considerations indicate the most economic conductor to be the one whose annual carrying charges (interest on investment, operation and maintenance costs, taxes, insurance, etc.) equal the annual cost of the power losses in the conductor. However, the drop in voltage and the heating of the line may be excessive. In such cases, the economy of the smaller conductor line may have to give way to one with larger conductors in order to produce tolerable voltage losses and heating conditions (Fig. 2.5). Again, the size of conductor may have to meet mechanical considerations and electrical losses because of the phenomena known as "skin effect" and corona (explained later in this chapter). Too thin a conductor may require additional supports because of its want of

TABLE 2.3 Clearances in Any Direction from Line Conductors to Supports and Guy Wires Attached to the Same Support

Rated Line Voltage (kV)	Suspension Insulators		Normal Clearance to Support	Minimum Clearance		
	No. Units*	Weight (lb)		NESC	To Support	To Guy Wires
34.5	3	38	1'7"	0'10"	1'0"	1'5"
46	3	38	1'7"	1'1"	1'0"	1'9"
69	4	48	2'1"	1'6"	1'3"	2'7"
115	7	78	3'6"	2'6"	2'2"	4'1"
138	8	88	4'2"	3'0"	2'9"	4'10"
161	10	108	5'0"	3'6"	3'6"	5'7"
230	12–14	138	6'10"	4'11"	5'0"	7'11"

*For average conditions tangent structure.

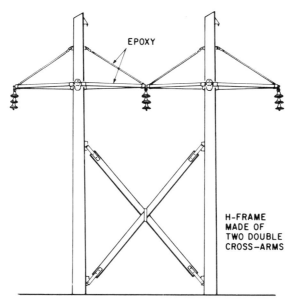

EPOXY

H-FRAME
MADE OF
TWO DOUBLE
CROSS-ARMS

Figure 2.4 Epoxy cross-arms.

mechanical strength, while on the other hand, large size conductors are difficult to handle and may impose larger than desirable strains on insulators. Here, one remedy is to subdivide the circuits into two or more circuits — which may add other advantages in reliability.

Self-inductance and Skin Effect

When a conductor carries an alternating current, it produces around itself a magnetic field whose intensity varies with the changing values of the alternating current (Fig. 2.6). This changing magnetic field cuts the conductor, inducing within it voltages which tend to retard the flow of current normally flowing in the conductor. This phenomenon is known as self-inductance or self-reactance. Within a conductor, this effect is more pronounced toward the center of the conductor. Hence, current flowing within that conductor will tend to flow more easily and consequently in greater part near the surface of the conductor. This is known as "skin effect," and is more pronounced as the operating voltages become larger. Where the conductor is stranded, the outer strands carry more current per strand than do the inner strands.

Corona

When a conductor carries a voltage exceeding a certain critical value, a halo-like glow, known as "corona," will appear on the surface of the conduc-

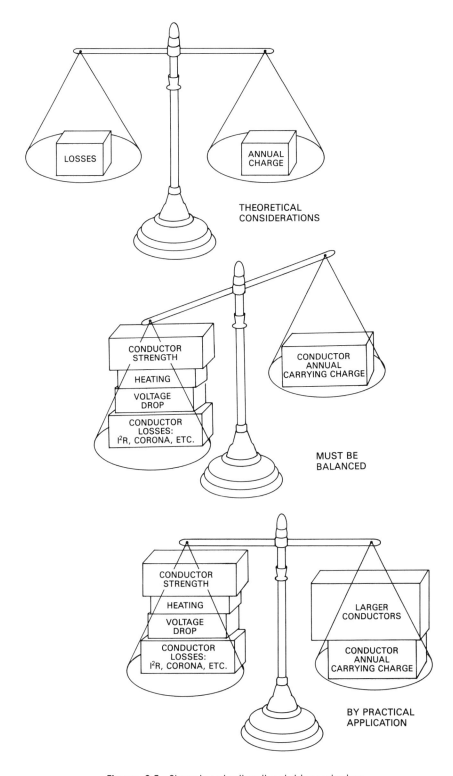

Figure 2.5 Steps in selecting the right conductors.

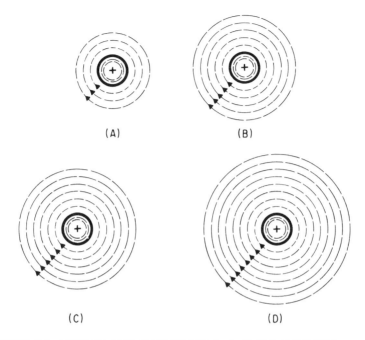

Figure 2.6 Magnetic field expanding (a, b, c, d) and contracting (d, c, b, a) about a conductor.

tor (Fig. 2.7). The amount of corona depends on the diameter of the conductor, the condition of the adjacent atmosphere, other nearby conductors, and condition of the surface of the conductor, such as dirt or roughness. This luminous effect is a discharge of electrical energy from the conductor into the atmosphere, where it is dissipated and represents a loss of power. If the distance between this conductor and other nearby conductors or structures is comparatively small, a sparkover may occur, triggering a short circuit at this point with consequent interruption or damage to the line.

Corona and skin effect may cause interference on communication circuits that parallel the transmission lines. Further, corona may also cause interference to local radio and television broadcasting.

Hollow Conductors

To lessen the effects of skin effect and corona, expanded conductors having hollow or partially hollow cores have been developed. Such conductors eliminate the center part of the conductor which is not fully used in carrying current because of the skin effect. Further, because a conductor of larger outside diameter results, the tendency for corona to appear is decreased.

A wire or conductor in which electricity may be flowing is surrounded

Figure 2.7 Corona effect.

not only by a magnetic field, but also by an electrostatic field. Electrostatic fields form generally in uniform patterns around a straight conductor and are also conductors of electricity. These patterns tend to become concentrated at points where the conductor is bent or at its edges or ends; the conducting tendency is also increased at these points. Since corona tends to appear at points where the electrostatic fields around a conductor are concentrated most, sharp bends or corners, and sharp points should be avoided. Corners or bends should be made gradual and smooth. Corona discharges may be greater, and actually observed, during period of rain. The rain drops accumulating on a conductor essentially change its shape; the clinging drops create relatively sharp pips on the conductor, encouraging the formation of corona at these points.

Corona Shields

To prevent corona flashover damage to insulators, particularly during inclement weather, shields (Fig. 2.8) or rings (Fig. 2.9) are provided at both the

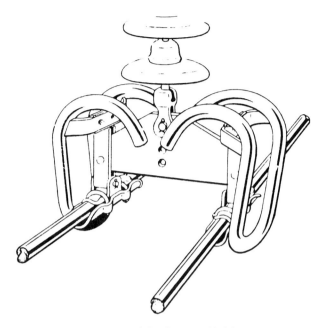

Figure 2.8 Corona shield.

conductor and the supporting end of the insulators. This furnishes a path for the flashover away from the surface of the insulators. The insulators are thereby protected from the shattering effect of the current which flows during the flashover.

Conductor Materials

Conductors can be made from various types of materials (Fig. 2.10). Although high conductivity (lesser resistance to the flow of electricity) combined with great strength and elasticity favors hard-drawn copper for use as a conductor, aluminum has an advantage because of the requirement for a large diameter to avoid corona. This, together with price advantages, makes it almost universally used in high-voltage transmission lines. The lower conductivity of aluminum, about 60 percent that of copper, results in a conductor having a diameter of 1.26 times that of a copper conductor of equal resistance and its lower weight makes it somewhat easier to handle. Its lower strength, some 70 percent that of copper, however, makes it less suitable for long spans. To overcome this shortcoming, aluminum cable, steel reinforced (ACSR) was developed. In this type of cable, aluminum conductors are wrapped around a steel cable which provides the strength necessary for longer spans. Later designs have replaced the inner steel cable with one made of a high-strength aluminum alloy.

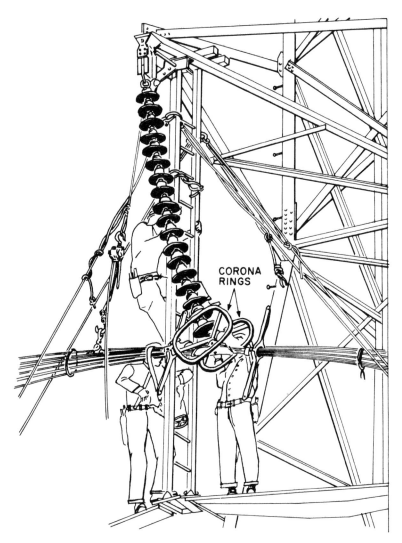

Figure 2.9 Corona rings.

For shorter lines designed to carry smaller amounts of power, conductors made of copper clad steel and aluminum clad steel, provide mechanical strength while furnishing relatively large diameters needed to mitigate skin and corona effects. For higher capacities, several of these clad wires may be used as strands of a larger conductor, or may be used as a core around which copper or aluminum strands may be laid to form a reinforced conductor.

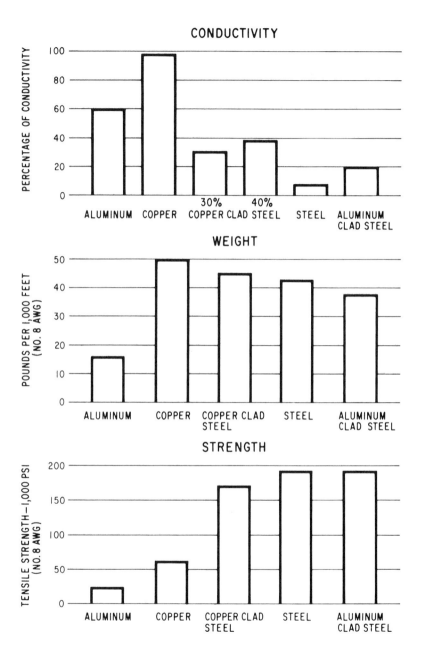

Figure 2.10 Comparison of various conductor materials according to conductivity, weight, and strength.

Steel conductors are occasionally used for extra-long spans, such as river crossings, where very high mechanical strengths may be required. Here, because the self-inductance is greater (because of the magnetic properties of steel) the skin effect may be great enough to require a larger cross-section for the steel conductor than for equivalent copper or aluminum conductors. The larger diameter serves to increase the value of critical voltage at which corona forms.

To minimize corona effects, the overall diameters of ACSR (and other) conductors have been expanded by introducing intermediate suitable fibrous material or forming air pockets between the outer aluminum strands and the inner steel core. Another means of minimizing corona losses is to replace the one conductor with a "bundle" of two, three, or four conductors, held in place by spacers, and separated from each other by suitable distances up to about 18 inches. Lower cost conductors may be used with this construction, though greater ice and wind loads may be experienced and greater sag for a given span may result.

Conductors for high-voltage transmission lines are handled very carefully to keep them from scraping on the ground during installation. Often special tension-stringing equipment is used that eliminates scratches or sharp edges forming on the conductor which may cause corona discharges. Aluminum is more subject to chemical attack and corrosion from substances found on the ground and in the atmosphere, than is copper.

Natural Hazards

Overhead transmission lines are exposed to attacks by the forces of nature. These include not only rain which affects the corona discharge on conductors, but also wind, and ice loading of conductors. Provision must therefore be made in the design of supporting structures for the effects of ice, wind, lightning, and other ravages of nature.

Aeolian Vibrations

Overhead conductors of transmission lines, where span lengths are relatively long, usually in open country, are subject to vibrations and movement produced by wind.

One type, known as aeolian vibrations (Fig. 2.11), is a regular high-frequency oscillation caused by the eddies behind a conductor produced by wind. The frequency of vibration depends on the size of the conductor, and wind velocity. To lessen the effect of these vibrations, dampers (masses of metal) are installed on the conductors, at node points (Fig. 2.12). Generally, true node points are impossible to be determined because of the varying factors producing the vibrations. In general, however, the dampers are located

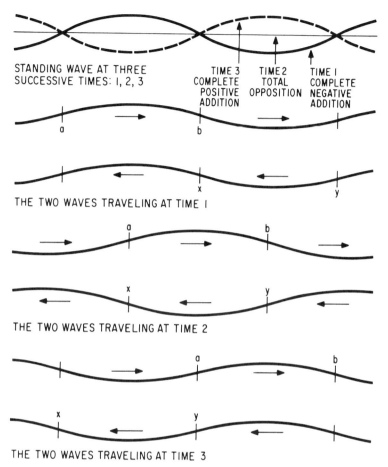

STANDING WAVE AT THREE
SUCCESSIVE TIMES: 1, 2, 3

TIME 3
COMPLETE
POSITIVE
ADDITION

TIME 2
TOTAL
OPPOSITION

TIME 1
COMPLETE
NEGATIVE
ADDITION

a   b

THE TWO WAVES TRAVELING AT TIME 1

a   b

x   y

THE TWO WAVES TRAVELING AT TIME 2

a   b

x   y

THE TWO WAVES TRAVELING AT TIME 3

NOTE: THE STANDING WAVE AT ANY TIME IS THE SUM OF THE TWO TRAVELING WAVES.

Figure 2.11 Standing waves on a conductor because of aeolian
vibrations.

near the towers at points calculated to give as much dampening as is practical.
Armor rods are also installed in aluminum conductors at the insulator clamps
to reduce the wearing effect of the vibrations in such conductors.

A self-damping type conductor is designed to reduce the aeolian vibration effect. This self-damping effect is achieved by making the shapes of the
conductor strands different. The outer strands are trapezoidal, while the inner
ones are round (Fig. 2.13). Relatively large clearances between these conductors permits motion to take place within the conductor layers which tends to
break up the vibration of the conductor caused by wind blowing against and
around it. With this conductor design, appropriate splicing and terminating
materials and procedures are necessary.

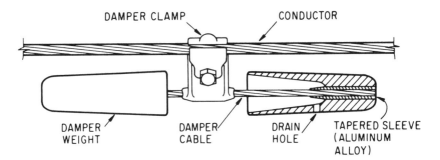

Figure 2.12 Dampers on a conductor.

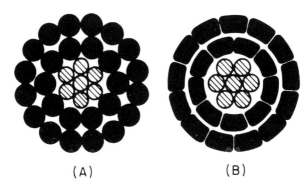

Figure 2.13 (a) Conventional ACSR; and (b) the self-damping conductor (SDC).

Galloping or Dancing Conductors

Another type of vibration, much more severe, is known as "galloping" or "dancing" conductors. The cause for this occurrence is not always known. However, ice forming on the conductors during freezing rain or sleet storms may sometimes assume the approximate shape of an airfoil. Wind blowing against it may cause the conductor to be lifted appreciably until it reaches a point where the conductor falls or is blown downward. Such a condition imposes extremely severe strains on both conductors and supporting structures, and may cause them to fail. Further, the increased sag of the conductors because of the ice load coupled with the nonrhythmic swaying may cause them to whip together causing flashovers and short circuits which may trip the circuits or burn down the conductors. Little can be done to correct this condition, fortunately rare, except to attempt to melt the ice from the conductors. This is sometimes done by overloading a circuit deliberately, either by connecting a "phantom" load to it, or by transferring loads from other circuits. Such overloads cause the conductors to heat and thus melt away the ice. This practice has been found successful in many instances, though sometimes uneven heating and dissipation of heat causes only portions of the conductors to become free of ice. If the circuits can be taken out of service, this is sometimes done, and the conductors are allowed to whip together with no danger of short-circuit and burndown.

Connectors and Splices

Mechanical connectors are almost universally used in splicing conductors. These may sometimes consist of sleeves or yokes which hold the conductors together by bolts. They may also be a sleeve into which the ends of the conductors are inserted and the sleeve crimped into them by hydraulically made indentations. These are referred to as compression type connectors.

These latter often have the indentations filled with solder and the whole splice polished, so that the tendency for corona to form about the splice is reduced. For ACSR conductors, two sleeves are used, an inner steel sleeve fitting over the steel core only, and an outer of aluminum fitting over the entire conductor (Fig. 2.14). For aluminum reinforced aluminum conductors, only one overall outer sleeve is used. Mechanical or hydraulic compressors are used to create the indentations on both the steel and aluminum sleeves which grip the conductors. Such indentations are usually started at the center of sleeve and proceed toward each end; the length of the sleeve may increase during compression.

Heat Detection

Splices or connections in the conductor are often sources of trouble as an improperly made, oxidized, loose connection, or broken strands may create hot spots in the line, which may cause failure. Usually, these are difficult to discover. A bolometer (a device used for measuring heat at a distance) equipped with a television camera and mounted on a truck or aircraft, which patrols the line, can successfully ferret out these potential sources of trouble before failure occurs.

Insulators

Insulators used for transmission lines are both of the suspension and pin or post types. In general, for tower lines and A-, H-, V-, or Y- (wood or metal) frame construction, standard insulator discs 10 inches in diameter and spaced $5\frac{3}{4}$ inches apart are used, the number required depending on the voltage of the line. Standards adapted by the Edison Electric Institute and National Electrical Manufacturers Association specify minimum numbers of discs in a string

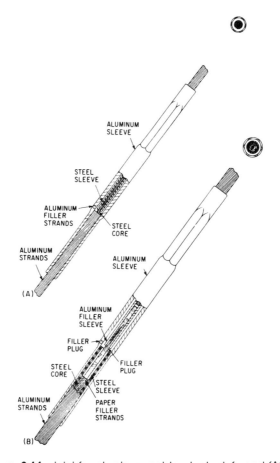

Figure 2.14 Joint for aluminum cable, steel reinforced (ACSR).

for certain line voltages: 4 for 69 kV, 7 for 138 kV, 12 for 230 kV, and 19 for 345 kV. Extra discs are added to provide safety factors, to allow for contamination and damage, and to compensate for the effect of altitude and metal structures as compared to wood on insulation requirements.

Suspension Insulators

The higher the voltage, the more insulation value is needed. Transmission lines operate at high voltages, 69,000 volts, for example. At these voltages the pin-type insulator becomes bulky and cumbersome. Besides, the pin which must hold it would have to be inordinately long and large. To meet the problem of insulators for these high voltages, the suspension insulator is used (Fig. 2.15).

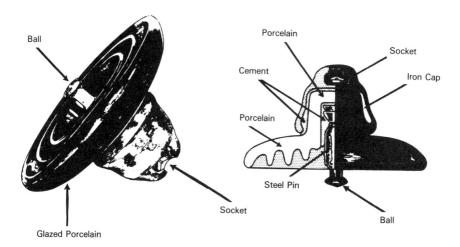

Figure 2.15 Ball and socket type suspension insulator.

The suspension insulator *hangs* from the cross-arm, as opposed to the pin insulator which *sits* on top of it. The line conductor is attached to its lower end. Because there is no pin problem, any distance between the cross-arm and the conductor may be obtained just by adding more insulators to the "string."

The entire unit of suspension insulators is called a string. How many insulators this string consists of depends on the voltage, prevailing weather conditions, the type of transmission construction, and the size of insulator used. It is important to note that in a string of suspension insulators one or more insulators can be replaced without replacing the whole string.

Insulator Assemblies—Disc Type

Disc insulators are usually made of porcelain, and designed so the porcelain is in compression (Fig. 2.16). Though they may be made in various sizes and shapes for specific purposes, the standard size disc mentioned above has received wide acceptance. Glass and some synthetic (usually epoxy) types are also in use. Where additional mechanical strength is required, as in very long spans, two or more strings of insulators may be paralleled to provide the additional strength.

The string of insulators may support a conductor vertically from a cross-arm or structure, or it may support it horizontally, a string in each direction, where strains are unusually great, as at turning points in the line [Fig. 2.17(a),(b)]. Such "strain" or "dead-end" insulator strings are also used on a long line to divide it into several sections so that, under unusual conditions such as storms, should a conductor fail, damage will be restricted to a small section. The unbalance caused by such a conductor failure may affect adjacent

Figure 2.16 Typical string suspension insulator.

spans in a domino-like effect. The installation of these dead-end strings prevents this effect from affecting the entire line.

Pin and Post Insulators

Pin-type (Fig. 2.18) and post-type [Fig. 2.19(a),(b)] insulators are generally confined to transmission lines of lower voltages, usually up to 69 kV. These may be mounted on cross-arms where, since the conductor is mounted on the top of such insulators, the pole height can be reduced by that amount. Post-type insulators may be mounted horizontally on poles in a so-called armless type of construction which has the advantages of better appearance and narrower right-of-way requirements.

Combinations of string insulators and horizontally mounted posts or insulated struts are used to control the position of the conductor, especially at angles in the line. Two strings of insulators, mounted from a cross-arm in a V position, are also used to hold a conductor in place away from towers or wood frame construction. These are referred to as vee-strings [Fig. 2.20(a),(b)].

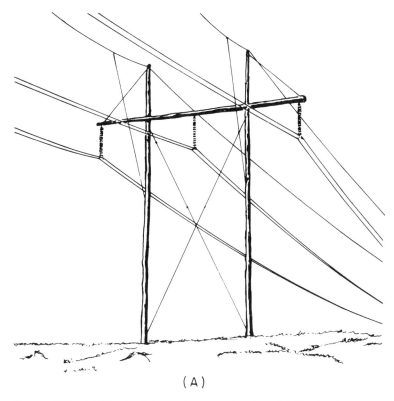

(A)

Figure 2.17 (a) Disc insulators—strain application; (b) Disc insulators—line (no strain) application.

Fittings

Fittings associated with the insulators are made of galvanized steel, malleable iron, or aluminum. The first two are used where strength requirements are high, the last for corrosion resistance and lessened corona effect because of the greater smoothness that can be attained. Aluminum coated steel fittings are also available.

Lightning

When lightning strikes at or near a transmission line, voltages are created in those lines greater than those at which the line normally operates. These "induced" voltages may be destructive if allowed to flashover to other lines or structures. Means are provided to "bleed" off these surges harmlessly. Arcing rings, mounted at one or both ends of the insulator string, permit the high-voltage surge to flashover between the ring and the structure or between these

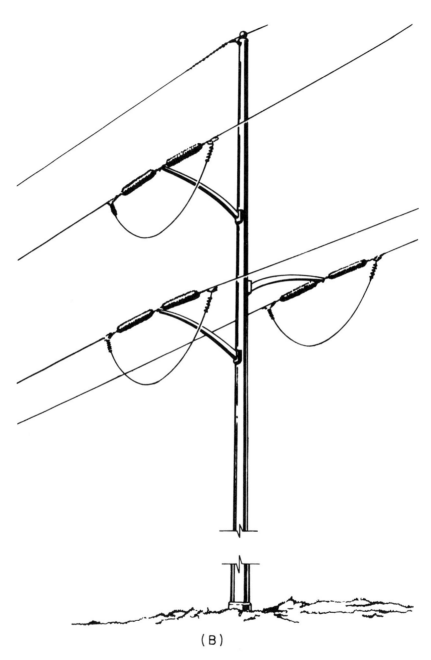

(B)

Figure 2.17(b) *(continued)*

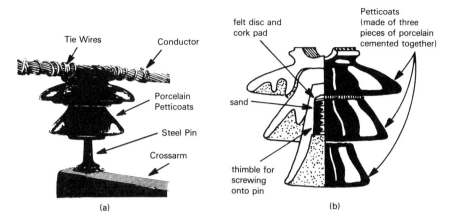

Figure 2.18 Pin-type insulator. (a) Mounted; (b) Cutaway view.

rings, away from the porcelain insulators, and the energy is passed over to ground where it is dissipated (Fig. 2.21).

Lightning or Surge Arresters

Lightning or surge arresters, usually a series of air gaps or special semiconducting materials, are also sometimes used to drain away the voltage surges. These are generally placed at the terminals of the lines to protect transformers, circuit breakers, and other equipment (at the substation). They may sometimes also be placed along the lines at strategic locations. The elementary lightning arrester (Fig. 2.22) consists of an air gap (horn air gap) in series with a resistive element. The overload voltage surge causes a spark which jumps across the air gap, passes through the resistive element (silicon-carbide, for example) which is usually a material that allows a low-resistance path for the high-voltage surge, but presents a high resistance to the flow of line energy. There are many different types of arresters, but they all have this one principle in common. There is always an air gap in series with a resistive element, and whatever the resistive (or valve) element is made of, it must act as a conductor for high-energy surges and also as an insulator toward the line energy. In other words, the lightning or surge arrester leads off only the surge energy. Afterwards, there is no chance of the operating line voltage being led into the ground.

The valve element in a resistor arrester (Fig. 2.23) consists of ceramic-like discs which act as conductors under high-voltage surges and present a high resistance to the line energy. In the arrester shown, the lightning current passes

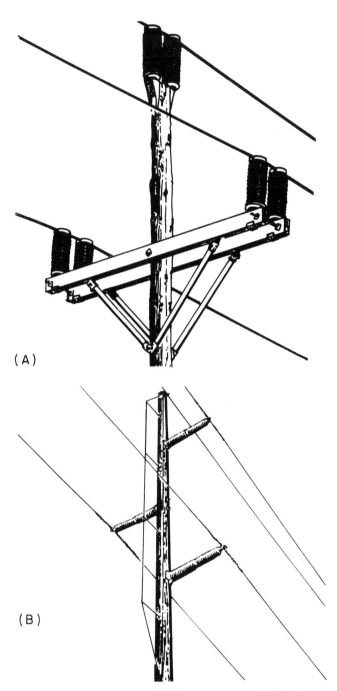

(A)

(B)

Figure 2.19 Post-type insulators. (a) Using cross-arms; (b) Armless construction.

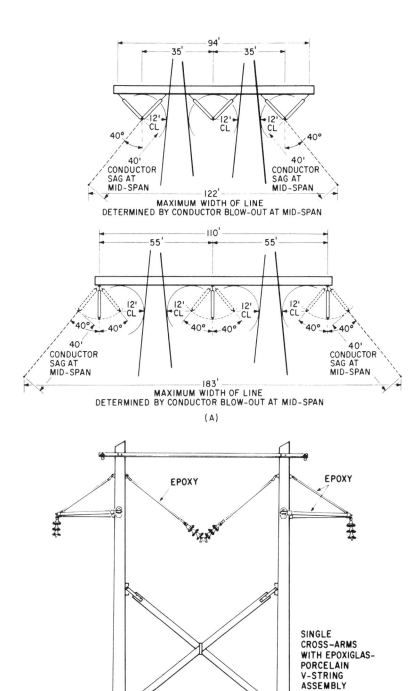

Figure 2.20 String insulator configuration. (a) Effect of v-string insulator configuration on the right-of-way requirement; (b) V-string insulators and insulated struts.

Figure 2.21 Arcing rings on the insulator.

through a series of by-pass gaps to the main gap and an element to ground. If the line energy attempts to follow the lightning energy, that line energy is made to flow through a series of coils which create a magnetic field strong enough to extinguish the arc of the lightning discharge. This extinguishing action is so rapid that it takes place in less than $\frac{1}{2}$ cycle of the line energy.

Shield or Ground Wires

Shielding these lines by means of a wire above them, which is directly connected to ground, is another form of protection from lightning. These wires will drain off the voltage surges from direct or nearby strikes of lightning [Fig. 2.24(a)]. Further, they cause the air adjacent to the lines to be drained of static electricity which sometimes serve to "attract" lightning discharges. This

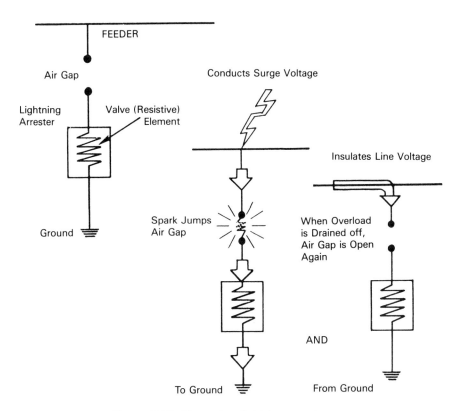

Figure 2.22 Elementary lightning arrester.

acts much the same as the ordinary lightning rod, and may be thought of as a continuous lightning rod, protecting the lines.

Either one or more wires may be mounted on the structures supporting the transmission line [Fig. 2.24(b)], depending on the type of structure, number of circuits, etc. It has been found that the installation of the overhead shield wires at an angle of 30° or less from vertical covering the lines generally results in the greatest protection of these lines. Wire for this purpose is generally of galvanized steel of small diameter, though copper clad steel is sometimes used to reduce electrical resistance while maintaining mechanical strength.

Counterpoise

To help this system drain off the lightning discharges, its resistance to the ground is kept low. This is further improved by burying rods or wire radially around the bases of the towers or supports, or burying a wire between them. This is known as a "counterpoise" [Fig. 2.25(a)], and its function is to dissipate

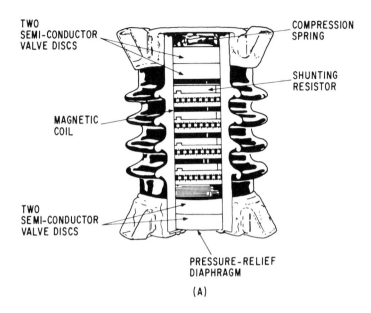

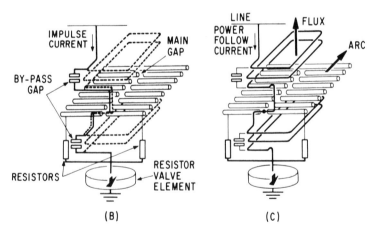

Figure 2.23 Resistor lightning arrester. (a) Cutaway view of resistor arrester; (b) Schematic diagram of resistor arrester showing path of lightning current in solid line; (c) Schematic diagram showing path of follow current in solid line.

the discharge over a larger area of ground, thereby lessening the resistance to the flow of these discharges. It is desirable, therefore, for the tower footing resistance to be made as low as possible; a value of 5 ohms or lower is generally sought. The question now arises as to how the value of tower footing resistance is measured and how it may be reduced.

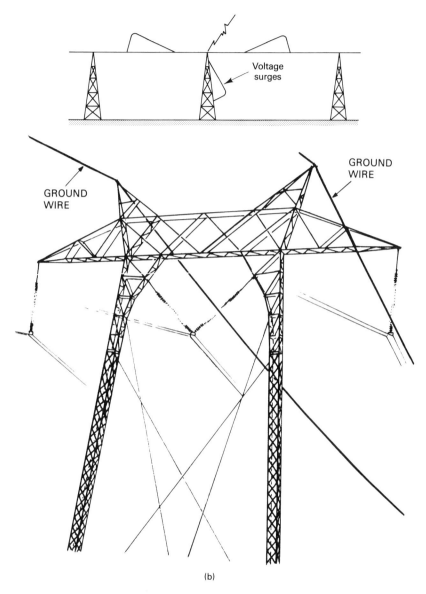

Figure 2.24 (a) Transmission-line ground wire struck by lightning; (b) Ground wires.

Methods of Measurement

There are three general methods used to measure the ohmic resistance from the tower to earth.

Fall of Potential Method.   In Figure 2.25(b), a known current, measured by the ammeter, flows in the ground between the tower and electrode B. The voltage drop, or fall of potential, between the tower and electrode A is measured with a voltmeter. The tower footing resistance (R_x) is then determined by the relation

$$\text{Footing Resistance } (R_x) = \frac{\text{Fall of potential}}{\text{Known current}}$$

The electrodes should be located sufficiently remote from each other to avoid proximity effect on each other.

Three-point method.   In Figure 2.25(c), three different resistance measurements are obtained: R_x in series with resistance of electrode R_A, R_x in series with the resistance of electrode R_B, and R_A in series with R_B. Three measurements are made, each one in turn, by using the voltmeter-ammeter method of measuring resistance. The footing resistance (R_x) is then computed from the equation

$$R_x = \frac{(R_x + R_A) + (R_x + R_B) - (R_A + R_B)}{2}$$

The electrode should have resistances of the same order of magnitude as that of the tower footing.

Ratio or Wheatstone Bridge Method.   In Figure 2.25(d), $R_x + R_B$ is first measured with the Wheatstone Bridge. Then

$$R_1 \text{ is to } (R_1 + R_2) \text{ as } R_x \text{ is to } (R_x + R_B)$$

or

$$\frac{R_1}{(R_1 + R_2)} = \frac{R_x}{(R_x + R_B)}$$

but

$$R_1 + R_2 = R_x + R_B$$

and

$$R_1 = R_x$$

$$R_x = \frac{R_1(R_x + R_B)}{R_1 + R_2)}$$

or

$$R_x = (R_x + R_B)\frac{R_x}{(R_x + R_B)}$$

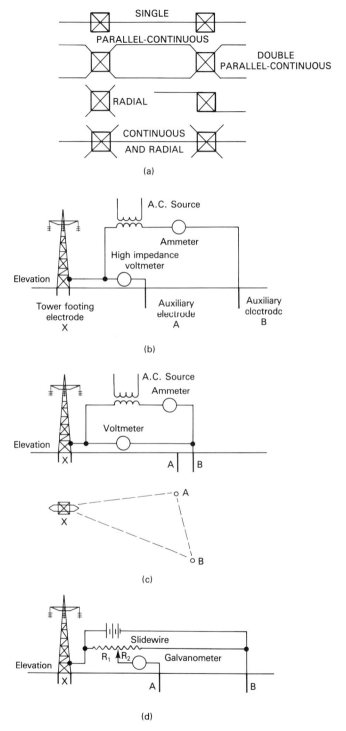

Figure 2.25 (a–d) Arrangement of counterpoises.

It should be realized that the resistance of an earth connection varies with many factors such as moisture, temperature, depth and diameter of electrodes, season of the year, and earth composition.

Methods of Reducing Tower Footing Resistance

Resistance of tower footings may be reduced: by installing additional conductor paths in the counterpoise; by driving additional ground rods and connecting them to the counterpoise; or by a combination of both.

In some regions, it is fairly easy to obtain tower footing resistances of 5 ohms or less; in other regions, it may be more difficult. Soil resistivity is the most important factor in determining the resistance of tower footings. Some typical values are shown in Table 2.4.

Surges

Similar "surges" will occur in transmission lines from switching operations, or from intermittent temporary short circuits. The first are created when a switch or circuit breaker is opened on interconnected lines, and the surge is created much as in a water system when a valve is suddenly closed. The latter may be caused by tree limbs, kites, or other foreign objects making temporary contacts. Rapid relaying is sometimes provided which takes the lines out of service before damage to them or other equipment can occur, and restores them to service by automatic reclosing devices called "reclosers," as soon as the transient disturbance is removed. Figure 2.26(a) is an oscillogram showing a typical example of a recloser operation. Notice that the first time it opens and

TABLE 2.4 Typical Values of Siol Resistivity*

Soil	Resistivity Range
Clay, moist	14–30
Swampy ground	10–100
Humus and loam	30–50
Sand below ground water level	60–130
Sandstone	120–70,000
Broken stone mixed with loam	200–350
Limestone	200–4,000
Dry earth	1,000–4,000
Dense rock	5,000–10,000
Chemically pure water	250,000
Tap water	1,000–12,000
Rain water	800
Sea water	0.01–1.0
Polluted river water	1–5

*In ohms per cubic meter.

closes, the action is instantaneous requiring only 1.6 cycles. The second time the action is delayed to 2 cycles, the third time to 6, and the fourth time to 5½ cycles. Then the recloser locks itself open and a worker must correct the fault and manually close the mechanism. The timing mechanism [Fig. 2.26(b)] which can effect this sort of action is a hydraulic system utilizing transformer oil as retarding fluid. With this device, the second, third, and fourth openings of the recloser can be set to any desired time span. To adjust the second and third openings, the time plate must be positioned. To designate whether the lockout should occur after two, three, or four operations, the cotter pin must be placed in the proper hole. To select the number of fast operations (one, two, or three) the roll pin must be positioned. If it is desirable to have all four operations instantaneous, the timing plate must be removed.

Maintenance

When such disturbances occur, it is usually customary to patrol the line to look for places of fault. Flashover marks, pitted conductors, chipped or stripped insulators, and fallen overhead ground or shield wires may be found. Maintenance and replacement of damaged units usually follow to prevent these relatively minor incidents from developing into major sources of trouble. Helicopters and aircraft have been used, as well as patrols by foot, horseback, and car.

Live-line Operation

Insulators, conductors, and other items may be handled while energized with live-line (hot stick) or bare-hand methods. Ingenious tools, mounted on the end of wood sticks are used in live-line maintenance. The wood acts as insulation. The worker fastens the stick to a conductor, after first disconnecting it from the insulator or insulators. Lashing the stick to the pole or tower keeps the energized conductor a safe distance away while insulators are changed or structure repairs made, or ground shield conductors are reinstalled. More sophisticated procedures are used to replace connectors, or wrap reinforcing rods around pitted conductors or broken strands. Care must be taken to see that workers always have sufficient insulation or space between them and energized parts of the line. For additional mechanical safety, conductors and other energized items are usually supported by one or more hot sticks, or by a hot stick and insulating rope.

Bare-hand Method

Methods have been developed to allow workers to work on conductors or other energized items as long as the platforms or buckets in which they are standing are insulated (Fig. 2.27). Work may also be done from a dolly which

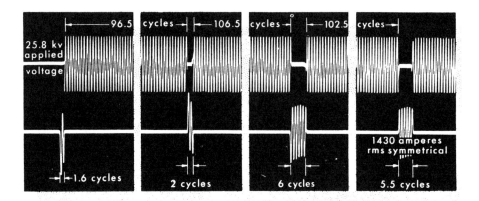

(A)

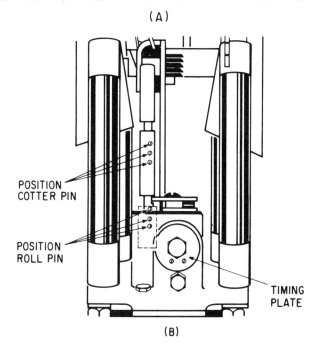

(B)

Figure 2.26 Automatic recloser operation. (a) Oscillogram showing redoser operation; (b) Timing mechanism for oil current recloser illustrating components for sequence adjustment.

rides on the conductor which is being worked on. In such an instance, no attempt is made to insulate the worker from the energized item. The worker now becomes energized together with the item. However, the support (or dolly) is insulated. Extreme caution is necessary in this method, as the insulated platform or bucket insulates the workers from a live conductor and

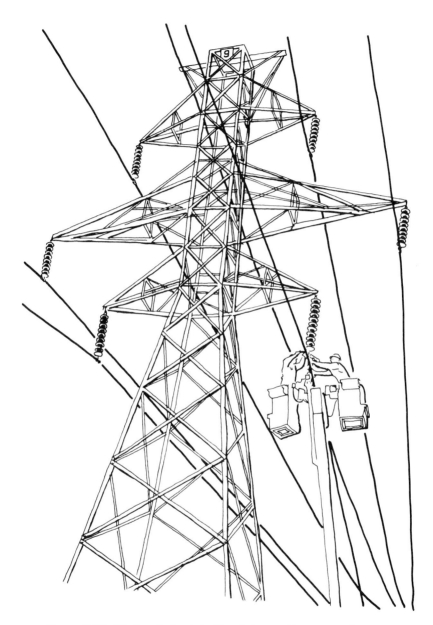

Figure 2.27 Workers in insulated buckets repairing power lines.

ground, but does not protect them when working on two or more live conductors between which high voltages may exist. Although known as "bare-hand" method, workers usually wear noninsulating work gloves.

REVIEW

☐ Structures for supporting overhead conductors are broadly classified as poles or towers. They may be made of natural wood: southern yellow pine, western red cedar, douglas fir, larch, and other species. They may be chemically treated, round shaped, tapered poles, or laminated round or square shaped poles where two or more layers of wood are glued together.

☐ Poles may also be of hollow tapered tubular design, made of steel or aluminum; or built up of flat metal members, latticed together, into a variety of cross sections. Reinforced concrete poles are used in special cases.

☐ Poles may be combines in A-frames, H-frames, and sometimes into V- or Y-type transmission structures (Fig. 2.1).

☐ Towers may be of several types: a tangent tower on which the conductor supported is essentially a straight line; a light-angle tower on which the conductor supported changes direction slightly, perhaps 5° or 10°; a medium-angle tower with support changes of 20° to 30°; and a heavy-angle tower which accommodates sharper turns.

☐ Generally, guys and anchors should be installed on dead ends, angles, long spans where pole strength may be exceeded, and at points of excessive unbalanced conductor tension (Fig. 2.2).

☐ The choice of supporting structures for transmission lines is influenced by many factors which, considered together, result in the greatest economy (Fig. 2.5).

☐ Both skin effect and corona may cause interference on communication circuits which may parallel the transmission lines. Corona may also cause interference to local radio and television broadcasting. To lessen the effects of skin effect and corona, hollow expanded conductors and corona rings are used in transmission lines.

☐ Cross-arms for towers are generally of galvanized steel or aluminum, and of wood for pole and A- or H-frame construction.

☐ In the ASCR type of cable, aluminum conductors are wrapped about a steel cable which provides strength.

☐ Insulators used for transmission lines are both of the suspension and pin or post types (Figs. 2.15, 2.17, 2.18).

☐ Long overhead conductors of transmission lines are subject to aeolian vibrations and galloping, or dancing conductors (Fig. 2.11).

☐ Arcing rings and lightning arresters protect transmission lines from damage caused by lightning. Shielding the lines by means of a wire above them, which is directly connected to ground, is another form of protection from lightning. Lower tower footing resistance is important [Fig. 2.25(a),(b),(c),(d)].

STUDY QUESTIONS

1. What are transmission line supports usually called? Of what material may they be made?
2. What type transmission structures may be made of wood or metal poles?
3. Into what types may towers be classified?
4. Where are guys and anchors generally installed?
5. What factors influence the choice of supporting structures?
6. What is meant by "skin effect"? What is corona? How may their effects be minimized?
7. Of what materials are overhead transmission conductors made?
8. What types of insulators are used on transmission lines? What are the advantages of each?
9. How do the forces of nature affect overhead transmission lines?
10. How are overhead transmission lines protected from lightning?

3

Underground
Construction

GENERAL CONCEPTS

Underground transmission lines may have some advantage of freedom from above ground weather and traffic problems, and thus experience fewer interruptions than overhead lines. However, there are a variety of failures that do affect cables. Interruptions underground may last from a few days to several weeks while the fault is found, the cable exposed, and repairs made under necessarily very exacting conditions. This time consumption compares to that of outages on overhead lines where fault location and repair are generally short-lived. Significant economic problems develop, therefore, because underground systems require that more facilities be available to attain the same level of reliability as that of the overhead systems. As underground transmission facilities are many times more costly than overhead, they appear feasible only in special areas, such as metropolitan centers, where towers or high poles in congested areas would be completely unacceptable.

Underground transmission cables may be laid directly in the ground in open fields or under suburban and rural roads. They may be drawn into ducts or installed in pipes in more or less densely populated urban and suburban areas, or laid in troughs in special cases such as under bridge structures or in tunnels [Fig. 3.1(a),(b),(c)]. The method of installation depends on the voltage and kind of cable, and the area in which it is installed.

64

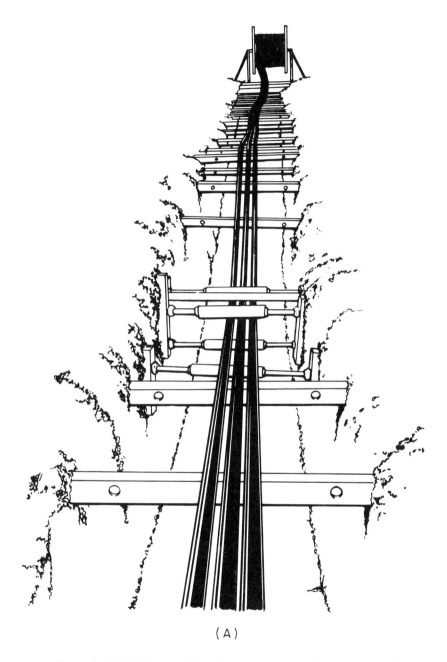

(A)

Figure 3.1 (a) Methods of installing underground transmission cables. (a) Buried directly in the ground; (b) Installed in ducts; (c) Installed in troughs.

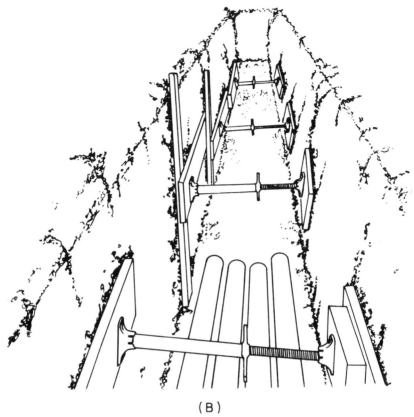

(B)

Figure 3.1(b) *(continued)*

CABLES

Tracking

High-voltage cables present particular problems, since the conductors of the cables are subjected to thermal expansion and contraction from the loading and unloading of current during daily or other periodic cycles. They initiate motion with respect to the insulation. Such motion tends to cause voids or pockets to form in the insulation. A void between the conductor and the insulation, or between the insulation and the grounded sheath, may also be formed because of faulty manufacture, bending too sharply during installation, or thermal expansion and contraction because of load cycling [Fig. 3.2(a)]. Such minute air pockets, under the electrostatic forces of the energized conductor, tend to ionize, that is, become conductors of electricity. This ionization of minute particles within the void causes corona discharge. The

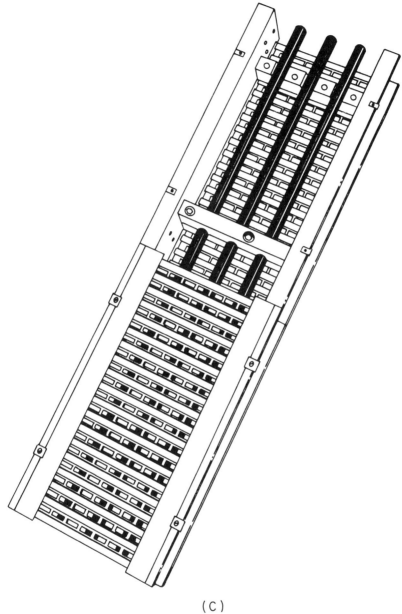

(C)

Figure 3.1(c) *(continued)*

corona discharge results in minute scorching and carbonizing of adjacent insulation. This is the beginning of "tracking" and the creation of ozone which damages the insulating value of most compounds [Fig. 3.2(b)]. When a sufficient number of these air pockets form, a tracking or charred path occurs where the insulation breaks down. Ultimately, the insulation is bridged with a carbonized track that is conductive and the cable fails [Fig. 3.2(c)]. Some high-voltage cables for underground transmission systems are shown in Figure 3.3.

Insulating Materials

Insulation for cables has largely been oil-impregnated paper, though more recent cables employ plastic materials. Some cables consist merely of paper wrapped around one or more conductors, and enclosed within a protective sheath. In others, the insulation wrapped around the conductors is subject to oil or gas under pressure. The gas used is usually nitrogen, but sulfur

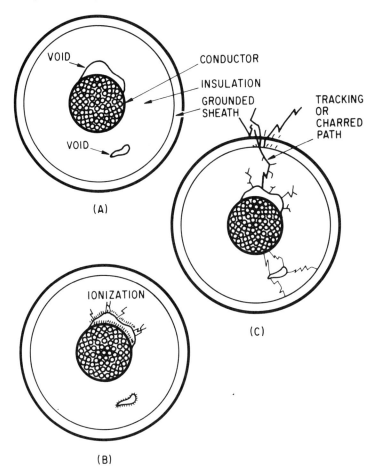

Figure 3.2 Tracking in a cable. (a) Void between the conductor and the insulation; (b) Corona discharge; (c) Cable failure.

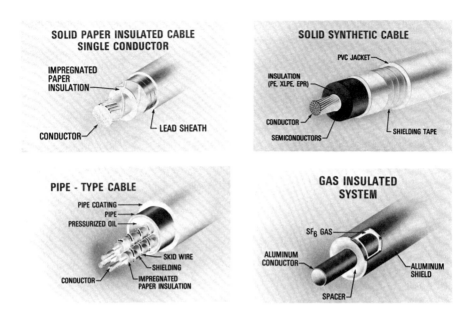

Figure 3.3 Some high-voltage cables for underground transmission systems.

hexafluoride (SF_6) has also been used. The theory here is that, when such voids occur in the insulation, the oil or gas under pressure fills the void and prevents ionization, tracking, and failure. Figure 3.4 shows the most suitable voltage ranges for paper-insulated cables.

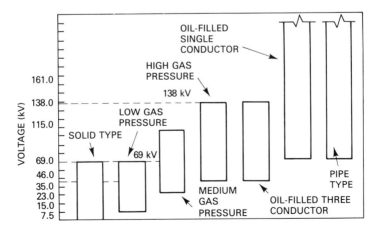

Figure 3.4 Recommended voltage ranges for various types of paper insulated power cables.

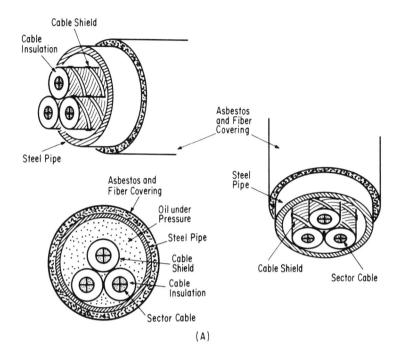

(A)

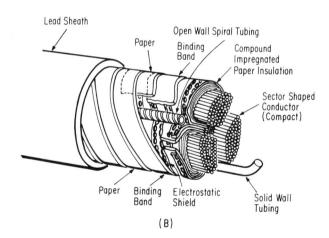

(B)

Figure 3.5 (a) Typical oil-filled and (b) gas-filled cables. (c) A comparison of power transmission capability of compressed gas, insulated cable, and oil-impregnated paper insulated cable. (*Courtesy of Edison Electric Institute.*)

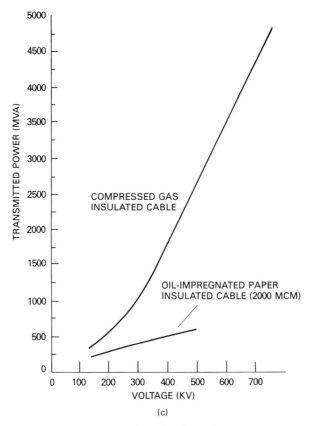

Figure 3.5 *(continued)*

Hollow Cables

In "hollow" cables, the oil or gas passage is located within the conductor or cable; in "pipe" cables, the oil or gas surrounds the conductors [Fig. 3.5(a),(b)]. Special accessories and auxiliary equipment to handle the oil or gas are needed with these latter-type cables. A comparison of power transmission capacity between gas and oil-impregnated insulated cables is shown in Figure 3.5(c).

Solid Insulation Cables

Solid insulated cables were formerly restricted to voltages of about 46 kV. However, with improved methods of manufacture and use of plastics, such as ethylene propylene, solid type cables of 69 kV and 138 kV have been built and put into operation.

Pipe Coverings

Hollow cables have been used in circuits operating from 69 kV to 350 kV. These installations require extraordinary care, since the pipe interior must be free of contamination. Evacuation before the introduction of the oil or gas therefore requires a high degree of vacuum and cleanliness. Special coverings on the pipe are used to avoid corrosion, electrolysis, and other damage to the pipe. Obviously such installations, while varying with the type of construction, the nature of the ground, road surface, transportation, and other facilities, are extremely expensive.

Cable Installation

Care should be exercised when installing cables by pulling them in a duct or pipe. This is especially necessary in the case of gas and oil-type cables. The stresses set up in the cables during this operation may cause damage to the conductors, the insulation, and the gas or oil paths. Outer armor wires, usually spiralled about a cable, may tighten about a cable so as to damage the insulation, and cut off the flow of gas or oil in that type cable. Some of the lengths of gas and oil cables pulled in at one time may measure in the thousands of feet. Dynamometers, instruments for measuring tension, are often employed during the pulling process to ascertain that allowable pulling stresses are not exceeded (Fig. 3.6).

Repair of Oil-filled Cables

While solid-type cables require relatively little maintenance and are comparatively easy to replace or repair, oil- and gas-filled cables present much greater difficulties. Oil or gas leaks occasionally occur and these must be found and repaired, generally without deenergizing the cable. Where oil-filled cables have to be replaced, whether because of failure, inadequacy, public improvement, or for other reasons, it is necessary to seal off the flow of oil on both sides of the section on which work is to be performed. This is done by freezing an oil slug from the outside of the cable or pipe by pouring liquid nitrogen until there is assurance the oil flow is stopped. These low temperatures are maintained by encasing the cable or pipe in ice and continued dripping of liquid nitrogen. The cable or pipe may then be cut, cables replaced, repaired, or rerouted, and in the case of pipe cables, the pipe assembly welded together again. Care is taken to reestablish the protective covering. Before reenergization, it is necessary not only to replace the oil in the affected section,

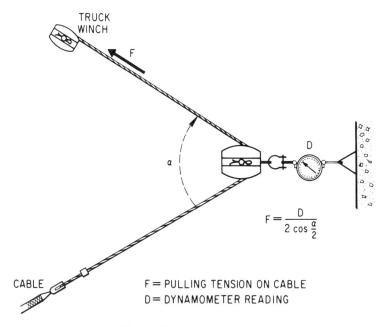

TRUCK
WINCH

F

α

D

$$F = \frac{D}{2 \cos \frac{\alpha}{2}}$$

CABLE   F = PULLING TENSION ON CABLE
               D = DYNAMOMETER READING

Figure 3.6 A dynamometer.

but to ascertain that all of the oil in the system is free from air, moisture, and other contaminants, a time-consuming and expensive process.

Figure 3.7 illustrates how freezing is employed. The method shown consists of an approximate 2-foot-long split lead sleeve with two stand pipes. The sleeve is opened sufficiently to pass round the cable. The gap is then closed and sealed along its longitudinal length with solder. The ends are then sealed to the cable jacket by the application of rubber tapes and reinforcing tapes such as hessian. The whole of the sleeve and the stand pipes are then insulated with asbestos or equivalent lagging either in tape or rope form.

Liquid nitrogen is poured into the freezing sleeve, via a funnel, to one of the stand pipes; the other acting as a vent for the exhaust of the nitrogen vapor.

Experience has shown that the time required to provide a solid blockage in a cable of approximately 2½-inch outside diameter is 30 minutes and for cables up to approximately 4 inches in diameter, 1 hour.

Repair of Gas-filled Cables

Repair of gas-filled cables generally follows the same procedures as for oil-filled cables except for one important difference. There is no oil to be frozen to stop the flow of oil out of the cable while it or a joint is being repaired. In gas cables, the gas is allowed to escape freely from between two

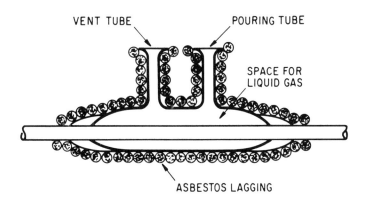

VENT TUBE   POURING TUBE

SPACE FOR
LIQUID GAS

ASBESTOS LAGGING

Figure 3.7 Freezing a section of an oil-filled cable to isolate a leak.

stop joints and replaced with clean dry gas when work is completed. Gas is fed at one of the stop joints and allowed to flow out at the stop joint on the other side of the repair. Gas is allowed to flow out for a brief period to ascertain no air or moisture remains in the cable before sealing at both stop joints.

Joints

In long installations, it is obvious that a gas or oil leak in either the hollow or pipe cable systems could lead to a long and costly decontamination process. To restrict the possibility of widespread contamination, such systems are usually sectionalized through the insertion of stop joints and semistop joints in the cable system. The number and types of such joints are usually dependent on the length of the line and its importance. The stop joint provides a physical barrier to both the conductor and the gas or oil flow system. These are used on long transmission lines and permit the line to be sectionalized so that the cables may be repaired or replaced without affecting the entire length of line. The semistop joint allows the conductor to pass through but imposes a physical barrier to the gas or oil-flow system (Fig. 3.8). Hence, in the event of a leak which would allow air or other contaminant to enter the cable, only a relatively small section of the line between the semistop joints is affected. In repairing gas-filled cables, the gas in one of the sections is completely replaced, and no operation similar to the oil freezing is required. Hence, both the time for repair and decontamination and the cost for such an incident are held down.

Direct Burial

When buried lead-sheathed or steel-armored cables are laid directly in the ground, precautions should be taken to prevent their being damaged.

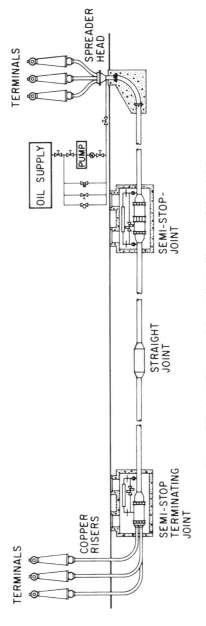

Figure 3.8 Isolating a cable section with a semi-stop joint.

Stones and other rough material should be kept away from the cable. The trench should be dug deeply enough to keep plows or diggers from damaging the cable. It should be dug wide enough to enable soft sandy fill to be packed around the cable to prevent bruises and cuts. This will also improve heat dissipation, described later in this chapter. A plank or other marker may be laid over the cable to act as a warning to workers and to protect the cable in the event the ground is disturbed after the cable is laid (Fig. 3.9).

Manholes

Manholes may or may not be provided, depending on the area of installation, length of cable runs, and other factors. They are usually required for pulling-in purposes and for splices both for original installation and for repair or replacement. Their dimensions and arrangement should be such that the cables are not subject to too short bending radii, but yet allow for cable movement under load or temperature differences so as not to put undue stress on the splice. Typically, for a 3-inch diameter cable, as shown in Figure 3.10, offset length L would be 66 inches, joint length, 30 inches, and the manhole length, 162 inches. In many instances, manholes have been eliminated and the entire installation is buried in the ground.

Soils

Underground cables, whether of the types buried directly in the ground or pipe types, are installed at depths below the frost line, usually two feet or more (Fig. 3.11). The soils in which they are buried have an important effect

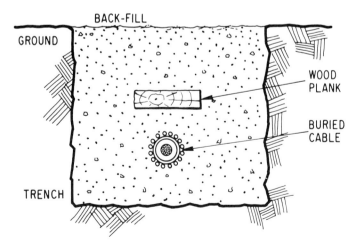

Figure 3.9 Using a creosoted wood plank to protect a buried cable.

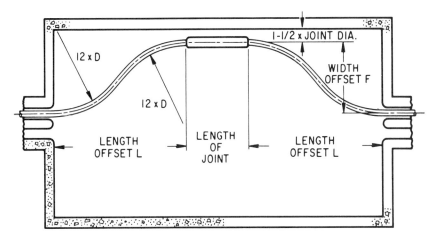

Figure 3.10 A typical manhole.

on the rated current-carrying capacity of these cables. The soils must provide adequate heat dissipation to prevent overheating and failure of the cables. Slag, ashes, or porous material are poor thermal conducting soils. Hot spots may occur along the cables, in which case the soil around that particular area may be replaced with earth or "thermal" sands having better heat-dissipating characteristics.

Splices

In splicing such cables, precautions of filling in the indentations on compression-type connectors, and smoothing the surfaces, to reduce or eliminate corona effects are taken, just as with overhead splices. The conductors

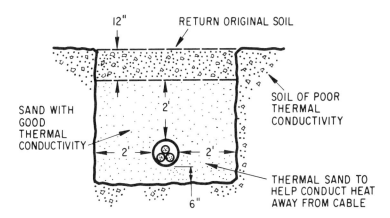

Figure 3.11 A cable buried in underground soil.

are usually of stranded copper, though they may be of aluminum. Hollow conductors lend themselves to lessening skin effect, similar to overhead type conductors.

Charging Currents

In underground transmission systems, the high voltage and the configuration of the conductor and metallic sheath produce a condenser effect, a phenomenon somewhat similar to the condition of clouds in thunderstorms that produce lightning. This is intensified as voltage levels increase, resulting in so-called charging currents that may exceed the cable thermal limits. This occurs if the length of the cable is sufficiently great, which makes the amount of current necessary to "charge" the cable rather large, thereby reducing its load-carrying capability. In instances where voltage levels are in the order of 345 kV or greater, total depletion of the load-carrying ability may occur after some 30 miles of transmission. Corrective equipment (shunt reactor) is available to compensate for this energy loss, but at extremely high cost. This cost must be added to the high cost of underground cable, the cost of splicing (which requires special skills), and other costs associated with underground installations. Figure 3.12 illustrates the charging current phenomenon. For a more technical explanation, refer to Chapter 6, Basic Electricity.

REVIEW

☐ Underground transmission cables may be (1) laid directly in the ground using armored cables, (2) drawn into ducts or installed in pipes, or (3) laid in troughs in special cases (Fig. 3.1). Insulation for cables has largely been oil-impregnated paper, though more recent cables employ plastic materials.

☐ In "hollow" cables, the oil or gas passage is located within the conductor or cable; in "pipe" cables, the oil or gas surrounds the conductors (Figs. 3.3 and 3.5).

☐ Dynamometers are often employed while pulling cables to assure that allowable pulling stresses are not exceeded (Fig. 3.6).

☐ Solid-type cables require relatively little maintenance and are comparatively easy to replace or repair, while oil- and gas-filled cables present much greater difficulties. Where oil-filled cables have to be replaced, whether because of failure, inadequacy, public improvement, or other reason, it is necessary to seal off the flow of oil on both sides of the section on which work is to be performed (Fig. 3.7).

☐ Stop joints are used in long installations to sectionalize the cable system to facilitate repair (Fig. 3.8).

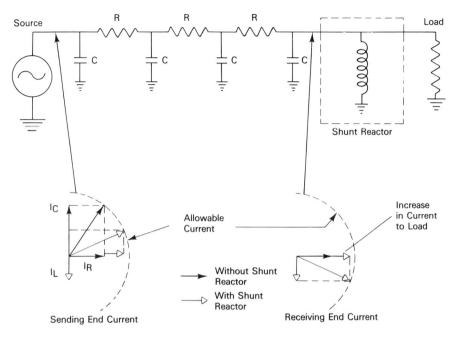

Figure 3.12 Simplified cable circuit (above) shows how power delivered to load is affected by a shunt reactor. In the absence of shunt reactors, this power decreases to zero at some critical length.

☐ Manholes are often required for pulling-in purposes and for splices both for original installation and for repair or replacement (Fig. 3.10).

☐ Underground cables, whether of the types buried directly in the ground or pipe types, are installed at depths below the frost line, usually 24 inches or more. The soils in which they are buried must provide adequate heat dissipation to prevent overheating and failure of the cables. If hot spots occur, that particular area may be replaced with earth or "thermal" sands having better heat-dissipating characteristics (Fig. 3.11).

☐ In splicing underground cables, precautions of filling in the indentations on compression-type connectors, and smoothing the surfaces to reduce or eliminate corona effects must be taken.

☐ In underground transmission systems, the high voltage and the configuration of the conductor and sheath produce a condenser action. This is intensified as voltage levels increase, resulting in so-called charging currents that may exceed the cable thermal limits (Fig. 3.12).

STUDY QUESTIONS

1. How may underground transmission cables be installed?

2. What do underground transmission cables consist of?
3. What may happen to insulation in underground cables, especially at the higher voltages?
4. What may the insulation of underground transmission cables consist of?
5. What is the difference between a "hollow" cable and a "pipe" cable?
6. What are stop joints and semistop joints, and where are they used?
7. How are sections of oil-filled cables replaced or repaired?
8. What is the function of manholes? What determines their design?
9. What effect does soil have on the cables which may be buried in it?
10. What precaution should be taken when splicing two conductors?

4

Substations

FUNCTIONS AND TYPES

Substations serve as sources of an energy supply for the local areas of distribution in which they are located. Their main functions are to receive energy transmitted at high voltage from the generating stations, reduce the voltage to a value appropriate for local use, and to provide facilities for switching (Fig. 4.1).

Substations have some additional functions. They provide points where safety devices may be installed to disconnect circuits or equipment in the event of trouble. Voltage on the outgoing distribution feeders can be regulated at a substation. In addition, a substation is a convenient place to make measurements to check the operation of various parts of the system.

Some substations are simply switching stations where different connections can be made between various transmission lines.

Some substations are entirely enclosed in buildings, while others are built entirely in the open (Fig. 4.2). In this latter type, the equipment is usually enclosed by a fence. Other substations have step-down transformers, high-voltage switches, oil circuit breakers and lightning arresters located just outside the substation building within which are located the relaying and metering facilities.

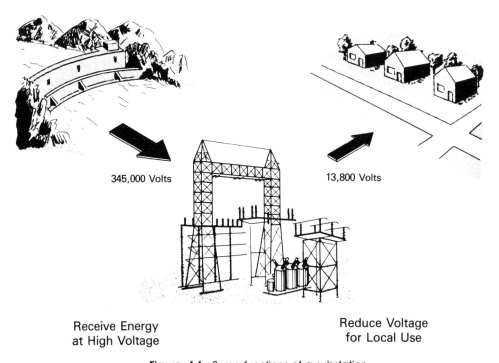

345,000 Volts   13,800 Volts

Receive Energy   Reduce Voltage
at High Voltage   for Local Use

Figure 4.1 Some functions of a substation.

Factors Influencing Location

Sites for substations are generally selected so that the stations will be as near as possible to the load center of the areas which they are intended to serve. Availability of land, cost, local zoning laws, future load growth, and taxes are just a few of the many factors which must be considered before a site is ultimately chosen.

Typical Features of a Transmission Substation

Substations usually have two or more incoming supply transmission lines for reliability. Many of these stations are operated automatically, with control circuitry back to an operating center. Such centers not only tell the operators the condition of the stations, but enable them to operate circuit breakers and other equipment remotely. These control circuits may be privately owned wires, public telephone circuits, or microwave installations. The substation often also serves as a convenient place where an overhead portion of a transmission line is connected to an underground portion.

The design of the substation arrangement should be such as to permit

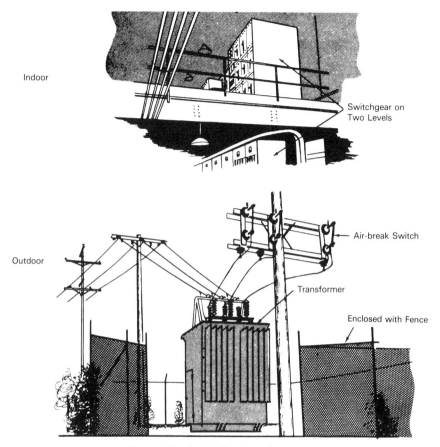

Indoor

Switchgear on
Two Levels

Outdoor

Air-break Switch

Transformer

Enclosed with Fence

Figure 4.2 Substation locations.

taking out of service several lines or units for operating or maintenance purposes without affecting the continuity of service. A typical substation arrangement is shown in Figure 4.3(a) together with standard symbols for equipment, shown in Figure 4.3(b).

Automatic Versus Manual Operation

Substations may have an operator in attendance part or all of the day, or they may be entirely unattended. In unattended substations, all equipment functions automatically, or may be operated by remote control from an attended substation, or from a control center (Fig. 4.4). In some unattended substations, the functioning of equipment will give an alarm at a remote station, and a "roving operator" will be dispatched to operate that station.

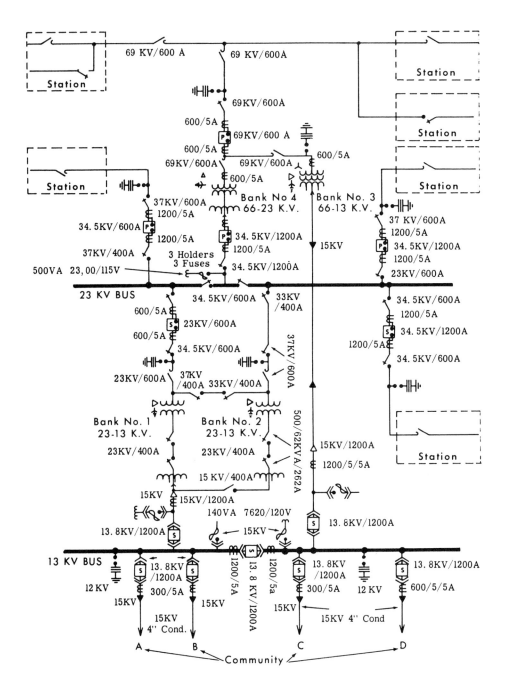

Figure 4.3 (a) Typical arrangement of a substation; (b) Standard symbols for equipment.

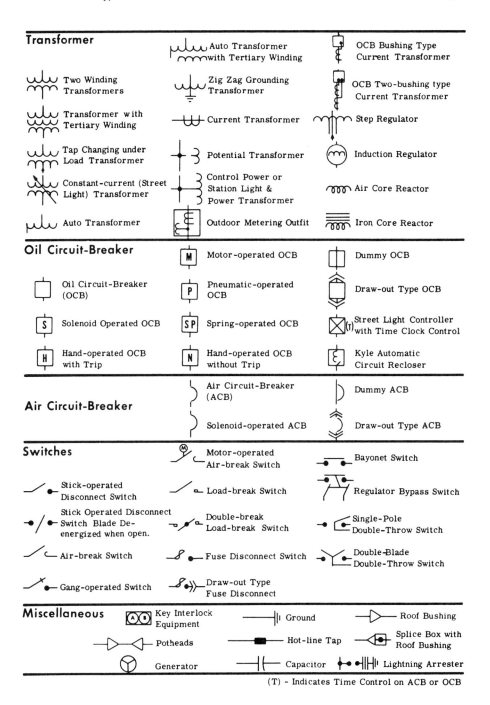

Figure 4.3(b) *(continued)*

RECORDING INSTRUMENTS
(Record Generation Capacity, Windpower,   INDICATING INSTRUMENTS
Voltage, Amperage, Wattage)   (Ammeter, Voltmeter, Wattmeter)

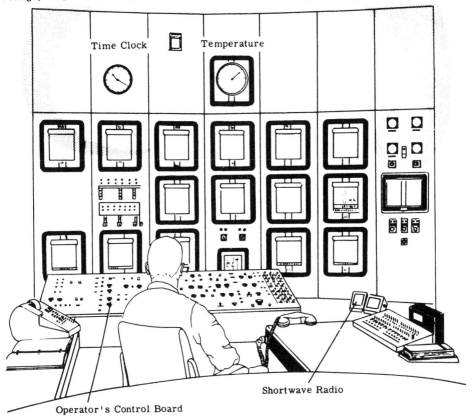

Figure 4.4 A control center which operates several substations by remote control.

Substation Equipment—The Transformer

The voltage of the incoming supply is changed to that of the outgoing subtransmission or distribution feeders by means of a transformer.

Fundamentally, a transformer consists of two or more windings placed on a common iron core (Fig. 4.5). All transformers have a primary winding and one or more secondary windings. The core of a transformer is made of laminated iron and links the coils of insulated wire that are wound around it. There is no electrical connection between the primary and the secondary; the coupling between them is through magnetic fields. This is why transformers are sometimes used for no other purpose than to isolate one circuit from

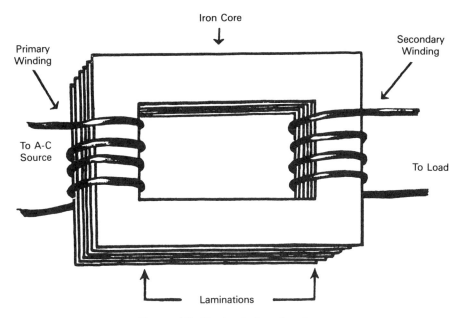

Figure 4.5 The basic transformer.

another electrically. When this is done, the transformer used for this purpose is called an "isolating" transformer.

The winding that is connected to the source of power is called the primary, and the winding connected to the load is called the secondary. The essential function of the conventional power transformer is to transfer power from the primary to the secondary with a minimum of losses. As we shall see, in the process of transferring energy from primary to secondary, the voltage delivered to the load may be made higher or lower than the primary voltage.

Step-up and Step-down Transformers

Transformers are wound to be as close to 100 percent efficient as possible. That is, all the power in the primary (as much as possible) is transferred to the secondary. This is done by selecting the proper core material, winding the primary and secondary close to each other, and a number of other careful designs.

Thus, assuming 100 percent transformer efficiency, we can then assume that the relationship between the primary and secondary voltages will be the same as the relationship between their turns. If the secondary has more turns than the primary we say that the transformer is operating as a step-up transformer; if the secondary has less turns than the primary we say that the transformer is operating as a step-down transformer.

Figure 4.6 is a representation of an autotransformer together with a conventional transformer. The autotransformer is somewhat different from the conventional transformer in that a portion of the primary and secondary is common, or makes use of the same turns. However, like all other transformers, the autotransformer does have the basic primary and secondary windings, although not physically isolated from each other.

Turns Ratio

Whether a transformer is of a step-up or step-down type the power in the primary is equal to the power in the secondary (Fig. 4.7). Thus, if the load draws 1000 watts, the product of voltage and current in the primary is also equal to 1000 watts. Another important principle is the fact that the primary and secondary voltages are in the same ratio as their turns. If the secondary has twice the turns of the primary, the secondary voltage will be twice as great as that of the primary.

Transformer Rating

The nameplate on a transformer gives all the pertinent information required for the proper operation and maintenance of the unit (Fig. 4.8). The capacity of a transformer (or any other piece of electrical equipment) is limited by the permissible temperature rise during operation. The heat generated in a

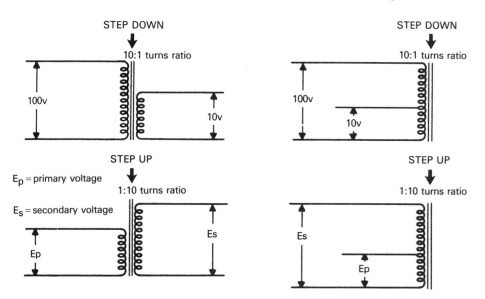

Figure 4.6 Comparison of a conventional transformer (left) and an autotransformer (right).

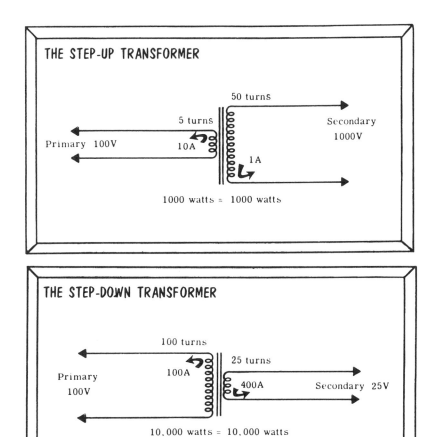

Figure 4.7 Step-up and step-down transformers.

transformer is determined by both the current and the voltage. Of more importance is the kilovolt-ampere (kV·A) rating of the transformer. This indicates the maximum power on which the transformer is designed to operate under normal conditions. Other information generally given on the nameplate is the phase (single-phase, three-phase, etc.), the primary and secondary voltages, the frequency, the permitted temperature rise, the cooling requirements — which include the number of gallons of fluid that the cooling tank may hold, and percent impedance (full load voltage drop, see Chapter 6). Primary and secondary currents may be stated at full load.

Depending upon the type of transformer and its special applications, there may be other types of identifications for various gauges, temperature indication, pressure, drains, and various valves.

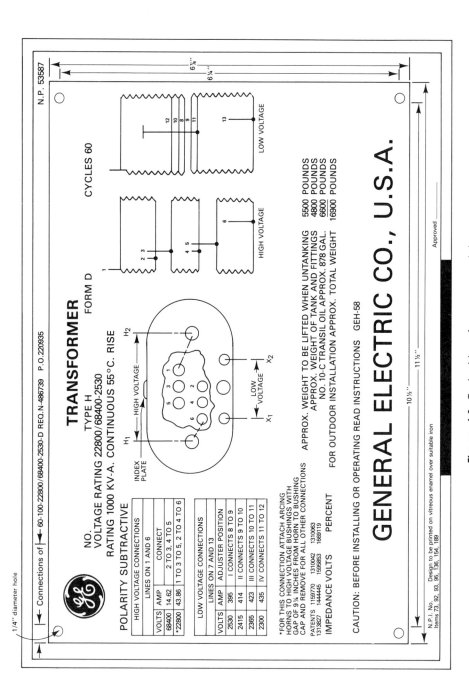

Figure 4.8 Typical transformer nameplate.

Thus, it can be seen that while the transformer consists primarily of a primary and a secondary winding, there are many other points to take into consideration when selecting a transformer for a particular use.

Methods of Transformer Cooling

The wasted energy in the form of heat generated in transformers because of unpreventable iron and copper losses must be carried away to prevent excessive rise of temperature and injury to the insulation around the conductors. The cooling method used must be capable of maintaining a sufficiently low-average temperature. It must also be capable of preventing an excessive rise in any portion of the transformer, and the formation of "hot spots." This is accomplished, for example, by submerging the core and coils of the transformer in oil, and allowing free circulation for the oil.

Since oil may be a fire hazard, sometimes, for reasons of safety, inert fluids, known as askarels, are used in place of the oil. These special fluids may be harmful to personnel handling them as well as to the varnishes generally applied to the insulation of the coils. Extreme care should be exercised in handling to prevent contact with eyes or open cuts and wounds.

Some of the oils and askarels in use have been found to contain polychlorinated biphenyl (PCB), an alleged cancer producing substance. Steps have been taken to eliminate the hazard by replacing the oils and askarels with cancer-free oils and other coolants. In some instances this is accomplished by replacing existing transformers; in other instances, the contaminating fluid may be drained at the site and, after several flushings with special contaminate-absorbing fluids that draw the PCB from the transformer core and parts, replaced with PCB-free oil or other fluids.

In clean dry locations, for indoor use, an open dry-type air cooled unit can be used. For outdoor (and indoor) use, a sealed dry-type unit can be employed [Fig. 4.9(a),(b)].

Some transformers (fluid filled or dry type) are cooled by other means: by forced air or air blast; by a combination of forced oil and forced air; and in some special applications, by water cooling.

Bus Bars

Bus bar (or bus, for short) is a term used for a main bar or conductor carrying an electric current to which many connections may be made.

Buses are merely convenient means of connecting switches and other equipment into various arrangements. The usual arrangement of connections in most substations permits working on almost any piece of equipment without interruption to incoming or outgoing feeders.

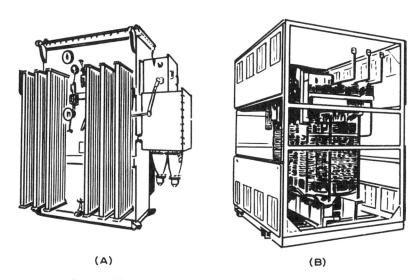

Figure 4.9 Oil-cooled (a), and air-cooled (b) transformers.

Some of the arrangements provide two buses to which the incoming or outgoing feeders and the principal equipment may be connected. One bus is usually called the "main" bus and the other "auxiliary" or "transfer" bus. The main bus may have a more elaborate system of instruments, relays, etc., associated with it. The switches that permit feeders or equipment to be connected to one bus or the other are usually called "selector" or "transfer" switches. As shown in Figure 4.10, bus bars come in a variety of sizes and shapes.

Substation Equipment—Regulators

A regulator is really a transformer with a variable ratio (Fig. 4.11). When the outgoing voltage becomes too high or too low for any reason, the apparatus automatically adjusts the ratio of transformation to bring the voltage back to the predetermined value. The adjustment in ratio is accomplished by tapping the windings, varying the ratio by connecting to the several taps. The unit is filled with oil and is cooled much in the same manner as a transformer. A panel mounted in front of the regulator contains the relays and the other equipment which control the operation of the regulator.

Substation Equipment—Circuit Breakers

Oil-circuit breakers (Fig. 4.12) are used to interrupt circuits while current is flowing through them. The making and breaking of contacts is done under

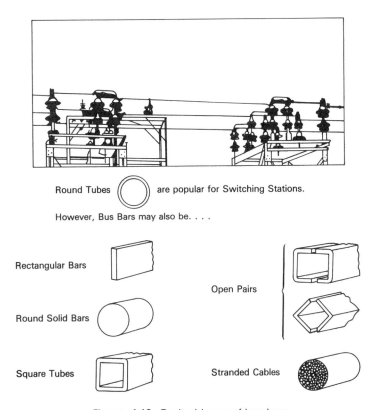

Round Tubes ◯ are popular for Switching Stations.

However, Bus Bars may also be. . . .

Rectangular Bars

Open Pairs

Round Solid Bars

Square Tubes

Stranded Cables

Figure 4.10 Typical types of bus bars.

oil. As explained previously, the oil serves to quench the arc when the circuit is opened. The operation of the breaker is very rapid when opening. As with the transformer, the high-voltage connections are made through bushings. Circuit breakers of this type are usually arranged for remote electrical control from a suitably located switchboard.

Some circuit breakers have no oil, but put out the arc by a blast of compressed air; these are called air-circuit breakers. Another type encloses the contacts in a vacuum or a gas (sulfur hexafluoride, SF_6) which tends to keep the arc from maintaining itself.

Substation Equipment—
Air Break and Disconnect Switches

Some switches are mounted on an outdoor steel structure called a rack [Fig. 4.13(a),(b)], while some may be mounted indoors on the switchboard panels. They are usually installed on both sides of a piece of equipment, to deenergize it effectively for maintenance.

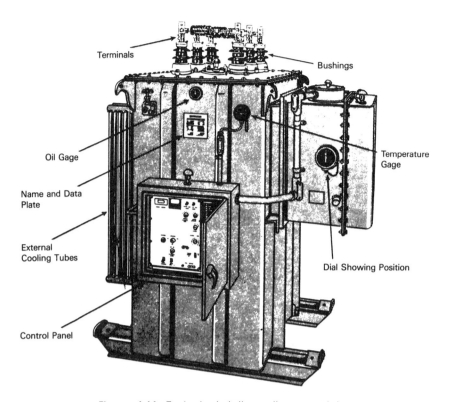

Figure 4.11 Typical substation voltage regulator.

Instrument Transformers

When values of current or voltage are large, or when it is desired to insulate the meter or relay from the circuit in which they are to operate, an instrument transformer is used.

In measuring current of high value, a current transformer (CT) is used (Fig. 4.14). The ratio of transformation is such that the high-current circuit, which in this case is the primary of the transformer, is reduced to a small current in the secondary connected to the ammeter or relay.

Similarly, a potential transformer (PT) (Fig. 4.15) has a fixed ratio of primary to secondary voltage. The secondary terminals are connected to the voltmeter or relay circuit.

Instrument transformers differ from power transformers in that they are of small capacity and are designed to maintain a higher degree of accuracy under varying load conditions.

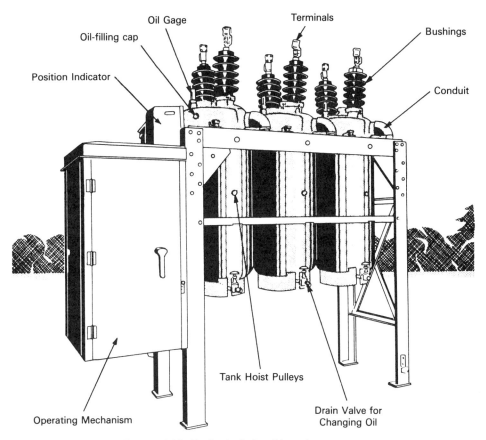

Figure 4.12 Typical oil-circuit breaker.

Substation Equipment—Relays

A relay is a low-powered device used to activate a high-powered device. In a transmission or distribution system, it is the job of relays to give the tripping commands to the right circuit breakers.

The protection of the lines and equipment is of paramount importance and is usually accomplished by the opening of circuit breakers automatically actuated by relays. In general, it is more important to provide protection for the components of a transmission system than on a distribution system. Greater blocks of load may be affected and resultant damage to lines and equipment may be more costly.

Relays are used to protect the feeders and the equipment from damage in

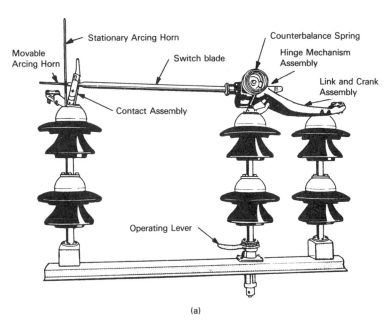

(a)

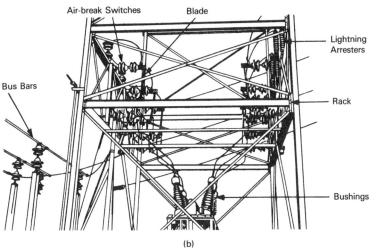

(b)

Figure 4.13 Air-break switches mounted on a substation rack.

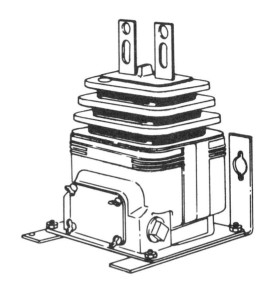

Figure 4.14 Typical current transformer.

the event of fault. In effect, these relays are measuring instruments, but
equipped with auxiliary contacts which operate when the quantities flowing
through them exceed or go below some predetermined value (Fig. 4.16). When
these contacts operate, they in turn actuate mechanisms which usually operate
switches or circuit breakers, or in the case of the regulator, operate the motor
to restore voltage to the desired level.

With the advent of miniaturization and of electronic gadgetry, including
silicon chips similar to those associated with computers, the operation of such
relays has become faster and more reliable.

Substation Bus Protection
Differential Relaying

In providing protection against faults on buses, current supplied to the
bus is measured against current flowing from it (Fig. 4.17). These should be
equal (with slight tolerances). When a fault develops on such a bus, this
balance is disturbed, and a relay will operate, usually clearing both incoming
and outgoing feeds from the bus. This is known as differential relaying.

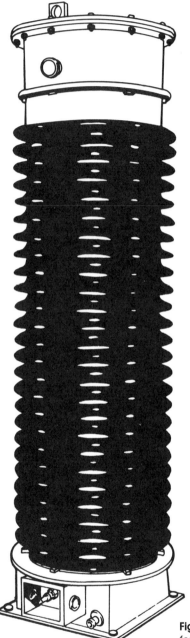

Figure 4.15 Typical potential transformer.

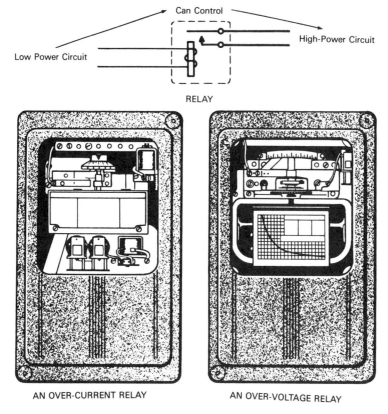

Figure 4.16 Typical substation relays.

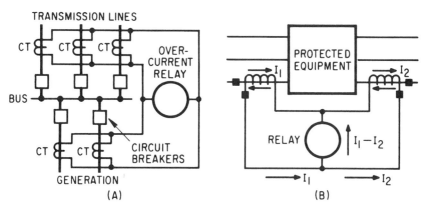

Figure 4.17 Relay protection. (a) Differential relaying is usually limited to equipment concentrated in one area due to the large numbers of control wires between sensing devices or current transformers. (b) A difference in current on either side of the protected equipment, due to a fault in the protected equipment, will be detected by current transformers and will pass through the relay actuating coil ($I_1 - I_2$). When $I_1 = I_2$, no current passes through the relay.

TRANSMISSION LINE PROTECTION

Relay Protection

When a fault occurs on a transmission line, from whatever cause, it is imperative that the line be deenergized quickly for safety reasons. This is accomplished by the opening of circuit breakers at each of its terminals, which are actuated by relays. The relays contain a movable element that is actuated by the current flowing through it. The movable element makes or breaks one or more sets of contacts, which in turn activate the mechanisms that operate the circuit breakers. The minimum value of the current that actuates the relays is predetermined for the basic *overcurrent* relay. When an additional element is added that will actuate the relay when the direction of current flow is opposite to the normal flow, the relay is termed a *directional* relay. When the movable element is actuated by a difference in the value of the current flowing into the line or equipment being protected and that away from it, the relay is called a *differential* relay. Refinements in the applications of the currents flowing in the relay result in relays sensitive to other features of line and equipment operation; these are discussed more fully in a companion book, *Electrical Transformers and Power Equipment.**

Applications of such protective relays are shown in Figure 4.18 for typical lines emanating from a generating station supplying one or more substations.

When a fault occurs on a transmission line near one of its terminals, the greater part of the fault current will flow through the circuit breaker of the nearest terminal. This will cause the circuit breaker to operate first, making the other end supply the fault current until the circuit breaker farthest from the fault operates somewhat later. The relatively much longer period in which fault current flows may cause severe damage. Hence, it is desirable that the circuit breakers at both ends of a faulted transmission line be opened simultaneously. This is especially so in the case of rather long transmission lines.

Pilot Protection

The simultaneous opening of circuit breakers at the terminals of transmission lines may be accomplished by means of a communication link between the circuit breakers involved.

This link may consist of physically separate pilot wires. Several methods of employing such pilot wires are shown in Figure 4.19(a),(b),(c),(d). Two schemes employing overcurrent relays at each end are shown in Figure 4.19(a)

*Anthony J. Pansini, *Electrical Transformers and Power Equipment,* Englewood Cliffs NJ: Prentice Hall (1988).

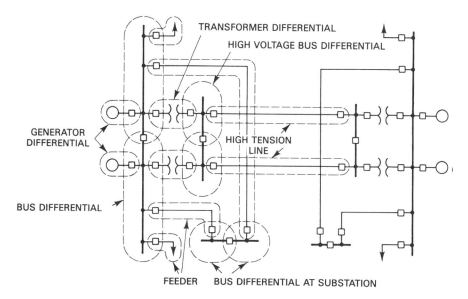

Figure 4.18 Protective zones for generators and outgoing high-voltage transmission lines. (*Courtesy of Westinghouse Electric Company.*)

and (b). A scheme employing differential relays at one end is shown in Figure 4.19(c). All of these employ three to six pilot wires. One scheme employing only two pilot wires, polyphase directional relays at each end, and a direct current source is shown in Figure 4.19(d).

Another more popular method requiring only two pilot wires employs an alternating current source and special type relays that combine the currents in each of the current transformers into a single-phase voltage that is compared to a similar quantity from the opposite end of the line; a simplified circuit is shown in Figure 4.20.

Often pilot wires are leased adding to the cost of such systems. Moreover, for positive and effective operation, the systems have a practical limit, usually of some ten miles. For longer lines, carrier pilot relaying and microwave relaying systems are employed.

Carrier Pilot Relaying

In this type of protection, the pilot wires are replaced by the transmission line conductors themselves, with a high-frequency current (50 to 200 kilocycles per second) superimposed on them. A simplified diagram and an installation at one end of the terminals are shown in Figure 4.21(a),(b). The carrier signal normally tends to operate the relays in such a manner as to keep the circuit breakers in a closed position. When the transmission line is faulted, the carrier

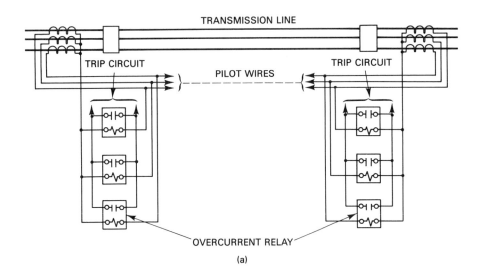

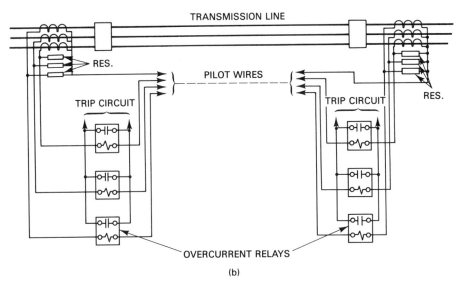

Figure 4.19 Pilot wire schemes: (a) Circulating-current scheme with load currents and through fault currents circulating over pilot wires; (b) Balanced-voltage scheme with load currents, and through fault currents, producing equal opposing voltages at line terminals; (c) Scheme using percentage differential relays; (d) Directional comparison scheme using direct current over a pair of wires (a–c connections omitted for simplicity). (*Courtesy of Westinghouse Electric Company.*)

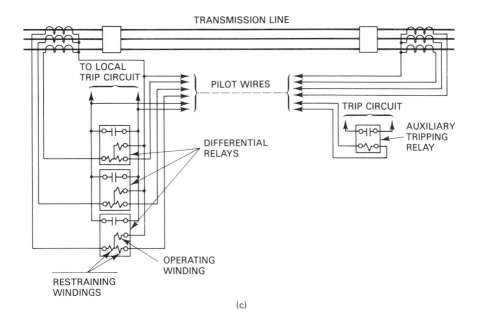

(c)

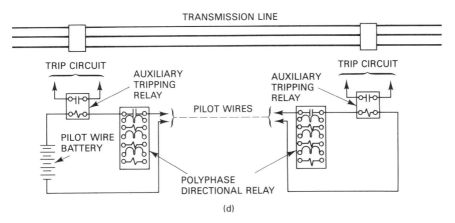

(d)

Figure 4.19 *(continued)*

signal is interrupted and the relays operate to open the circuit breakers. The system functions positively and effectively over several hundred miles. The carrier channel may also be used for other purposes; e.g., telemetering and supervisory controls that operate on coded impulses, for telephone communication and other nonrelated business.

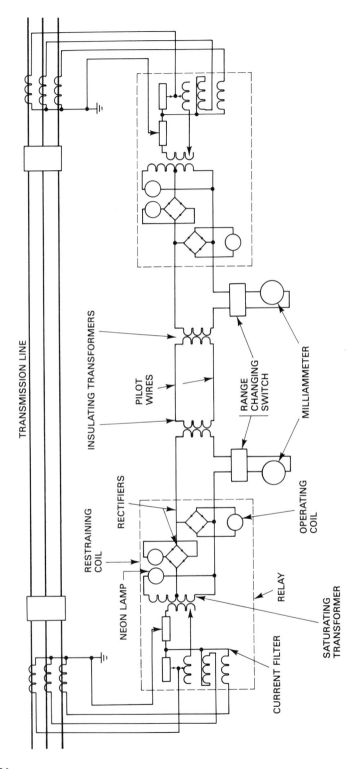

Figure 4.20 Alternating-current pilot wire scheme using special relays. (*Courtesy of Westinghouse Electric Company.*)

TRANSMISSION LINE

INSULATING TRANSFORMERS

PILOT WIRES

RANGE CHANGING SWITCH

MILLIAMMETER

RECTIFIERS

RESTRAINING COIL

NEON LAMP

OPERATING COIL

RELAY

SATURATING TRANSFORMER

CURRENT FILTER

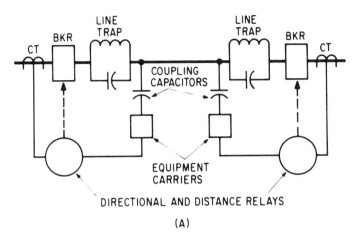

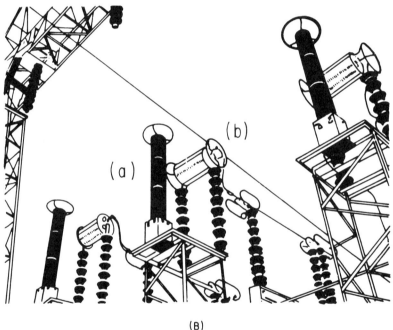

(B)

Figure 4.21 Carrier relaying. (a) Carrier equipment; (b) Illustration depicting installation of coupling capacitor (1) and line trap (2) on a 345KV transmission line.

Microwave Relaying

In microwave relaying, the pilot systems are transmitted over microwave radio channels. This system is not subjected to line faults, and it may also accommodate several other separate functions. Because the transmissions are affected by line-of-sight limitations, several units may be required for long and tortuous lines; the intermediate units receive their signals and retransmit them to the next unit (referred to as *relay stations*).

Generally, the pilot schemes operate normally to keep the circuit breakers closed. When the line is faulted, the signals are interrupted and the relays operate to open the circuit breakers. Similarly, if the pilot systems, including carrier and microwave systems, fail for any reason, the circuit breakers open to deenergize the lines and the transmission system "fails safe."

Ground Relay

Other schemes measure the flow of so-called ground current. In a transmission line, the currents flowing in each of the conductors are usually fairly well balanced in magnitude, so that the return or ground conductor carries little or no current. By measuring this directly, or by measuring each conductor and determining the difference, this ground current can be made to actuate relays when it exceeds certain predetermined values.

Other more sophisticated schemes are sometimes used, but generally they employ one or more of the described basic ideas.

Figure 4.22 shows a relay using ground fault detection, and illustrates the ability of this protection scheme to discriminate between load current and fault current.

Voltage Surges

In considering the effects of voltage surges, whether from lightning, switching, or fault conditions, it is necessary to consider not only the transmission lines themselves, but also the apparatus and equipment which may be connected to them. These include switches, circuit breakers, transformers, generators, buses, regulators, and any other device which may be connected to them. Generally, these are situated at generating stations and substations. Much of this equipment is similar to and operates in the same manner as that found in distribution substations. There are several, however, which are of greater importance for the transmission system.

Figure 4.23 is a graphical representation of a traveling wave on a transmission line showing two of several possibilities of voltage surges occurring at points of discontinuity, such as an open switch, a transformer bank, or change of overhead to underground. In (a), the characteristics of the point of discontinuity are such that the reflected wave is superimposed in the original surge voltage wave and the crest voltage is double the original value at the point of discontinuity. In (b), the characteristics of the point of discontinuity are such that the reflected wave is subtracted from the original wave.

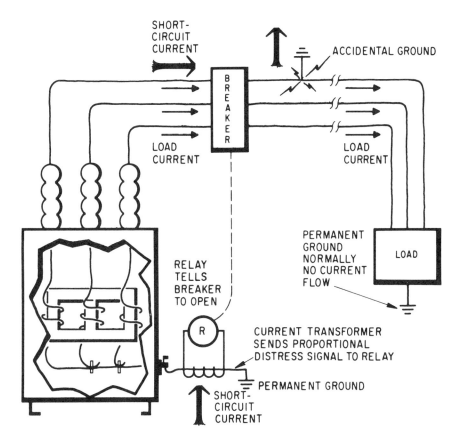

Figure 4.22 Ground relaying.

Basic Insulation Level (BIL)

Since the operating voltages in a transmission system are relatively high, special attention is given to the insulation associated with the several parts of the system, both in the lines as well as in the stations. The insulation here has to withstand not only the normally applied operating voltages, but the surge voltages explained later (Fig. 4.24). Since such insulation is expensive, it is desirable not to "over" insulate any portion of the system unnecessarily. Hence, the insulation of the several components of the entire system must be coordinated, and a level of insulation to be provided decided upon. Levels of insulation which will safely sustain the surge voltages, known as basic insulation levels (BIL), have been set up for electrical apparatus, such as transformers, circuit breakers, and switches. These minimum levels designated may

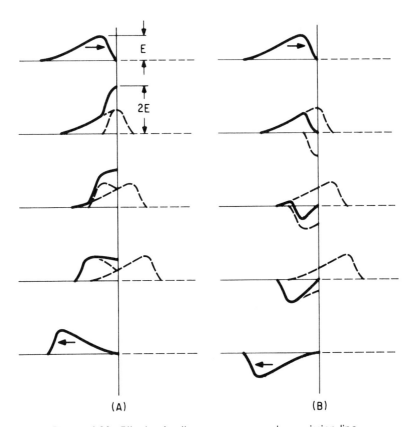

Figure 4.23 Effects of voltage surges on a transmission line.

be from 2 to 3½ times the normal operating voltage, depending on the degree of reliability desired. The minimum insulation level should prevail under wet conditions, and in general, should be the same for line insulation as for apparatus insulation.

At some point, usually the bushing, the insulation value is made deliberately at the lowest value so that, should failure occur, it will occur at a point that is readily accessible for repair or replacement.

Short-Circuit Duty

Again, because of the high voltages employed, when a fault or short circuit occurs on a transmission line, the amount of current which will flow will be inordinately great. This current flows, not only through the transmission line conductors, but through all the apparatus connected to it. This high magnitude current produces magnetic fields of great intensity with corresponding great forces. These forces tend to pull conductors apart and cause

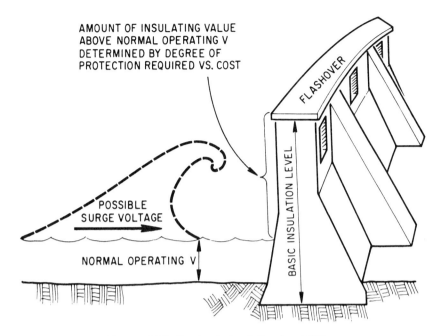

AMOUNT OF INSULATING VALUE
ABOVE NORMAL OPERATING V
DETERMINED BY DEGREE OF
PROTECTION REQUIRED VS. COST

FLASHOVER

POSSIBLE
SURGE VOLTAGE

BASIC INSULATION LEVEL

NORMAL OPERATING V

Figure 4.24 The basic insulation level (BIL).

damage to lines and equipment. Hence, equipment, particularly circuit breakers, must be built rugged enough to withstand these disruptive forces; the measure of this ruggedness is referred to as the interrupting capacity or "short-circuit duty" and is generally expressed in kV·A (thousand volt amperes) or mV·A (million volt amperes). Hence, circuit breakers and other equipment are rated not only for their normal voltage and current carrying ability, but also their so-called short-circuit duty; e.g., 69 kV, 500 amperes, 500 mV·A. As transmission grids become larger, with more generation interconnected, these fault currents and associated short-circuit duty become larger, making equipment more expensive.

Stability

When faults occur on a transmission grid supplied by two or more generators (Fig. 4.25), the current flow to the fault will be proportioned to the distance (electrically) of the several generators. Thus, the generator closest to the fault will supply the greatest share [Fig. 4.25(a)]. As these heavy currents are imposed on the several generators, it will cause them to slow down, but not equally; again, the one supplying the greatest share of current will slow down the most. Hence, these generators will no longer operate "in step," but the one which slowed down least will attempt now to supply the other generators connected to the grid which are now "bucking" it. This will cause it to slow

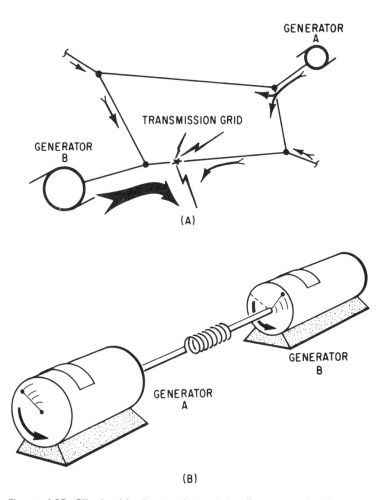

Figure 4.25 Effects of faults on a transmission line connected to more than one generator. (a) Short circuit in a transmission grid. Generator B supplies a major portion of the short circuit current. This causes generator B to slow down. (b) Generator A tries to assist. The effect is similar to a spring connection between generators, tending to keep them in step.

down, and the process reversed. A rocking back and forth effect between generators will ensue. The generators then act very much as if they are connected together mechanically through a spring [see Fig. 4.25(b)].

If the fault is removed in time, this rocking effect will subside and the generators will snap back into step. If not, the effect will increase progressively until the current supplied by some generator exceeds its protective relay setting, and it will be disconnected from the grid. If the fault persists, similar effects will cascade to other generators until all the units are disconnected and a "blackout" results. This effect is generally referred to as lack of stability or an unstable condition. Settings on protective relays and devices are designed to switch out the faulted section of the system as rapidly as possible, restoring the rest of the system to normal.

COGENERATION

Large industrial or commercial consumers who may have an excess of steam energy that can be converted into electricity, or who may generate their own electricity and have excess capacity, may sell their excess to utility companies. This is accomplished by interconnecting their installation to the utility transmission lines, usually through a substation-type installation. They must not only provide the necessary equipment and protective devices, but must coordinate them with the utility to which they may be connected. The power they put out may also be wheeled to other consumers or utilities. In any case, this portion of their operation must be placed under the control of the utility or pool system operator. Such installations may be considered as another of the transmission substations.

REVIEW

☐ Substations receive energy transmitted at high voltage from the generating stations, reduce the voltage to a value appropriate for local use, and provide facilities for switching. They also provide points where safety devices may be installed to disconnect circuits or equipment in the event of trouble.

☐ Substations are generally located so that they will be as near as possible to the load center of the areas which they are intended to serve. Availability of land, cost, local zoning laws, future load growth, and taxes are some of the many factors which must be considered.

☐ Substations usually have two or more incoming supply transmission lines for reliability. Many of these stations are operated automatically, with control circuitry back to an operating center. Substations may have an operator in attendance part or all of the day, or they may be entirely unattended. In some unattended substations, the functioning of equipment will give an alarm at a remote station, and a "roving operator" will be dispatched to operate that station.

☐ The voltage of the incoming supply is changed to that of the outgoing subtransmission or distribution feeders by means of a transformer. The nameplate on a transformer gives all the pertinent information required for the proper operation and maintenance of the unit (Fig. 4.8).

☐ A bus bar is a main bar, or conductor, carrying an electric current to which many connections are made (Fig. 4.10).

☐ A regulator is a transformer with a variable ratio that maintains a transmission voltage at the specified level (Fig. 4.11).

☐ Circuit breakers (Fig. 4.12) allow interrupting a circuit while current is flowing through it. Oil-circuit breakers are the most common type, but another type, called an air-circuit breaker, puts out the arc by a blast of compressed air. Another type has its contacts enclosed in a vacuum or a gas, which tends to keep the arc from maintaining itself.

☐ Instrument transformers are used to measure large currents or voltages and to insulate a meter or relay from the circuit in which they are to operate (Figs. 4.14 and 4.15).

☐ Relays provide protection for the components of a transmission system by providing tripping commands to the circuit breaker or breakers in an overloaded circuit (Fig. 4.16).

☐ Differential relaying provides protection against faults on buses by comparing current supplied to the bus with current flowing from the bus. The currents should be approximately equal. A fault on a bus creates an imbalance which triggers a relay to clear both incoming and outgoing feeds from the bus (Figs. 4.17 and 4.18).

☐ In carrier relaying, input and output currents are measured at both ends of a line, and pilot wire, microwave transmission, or other means are used to transmit these quantities to relays at both ends to clear the lines in trouble (Figs. 4.19a,b,c, and d and 4.20). Sometimes, the conductors themselves are used to transmit these signals (Fig. 4.21a&b).

☐ In ground relaying, ground current which is produced when the current flowing in the conductors is unequal, can be made to actuate relays when it exceeds certain predetermined values (Fig. 4.22).

☐ Basic insulation level (BIL) is the level of insulation that will safely sustain surge voltages. BIL must be considered when selecting devices such as transformers, circuit breakers, and switches. Minimum levels designated may be from 2 to 3½ times the normal operating voltage, depending on the degree of reliability desired (Fig. 4.24).

STUDY QUESTIONS

1. What is the function of a transmission substation? What other purposes may it serve?

2. What are some features of transmission substations?

3. What important factor influences the location of a transmission substation? What other factors are also considered?

4. What pieces of equipment may be found in a transmission substation?

5. What is the function of a substation transformer? Of a regulator?

6. What is the function of a circuit breaker? What types are there? How are they rated? Show a typical rating.

7. Why is it more important to provide protection for components of a transmission system than for a distribution system? How is this accomplished?

8. What are the functions of air-break switches and disconnects?

9. What is a relay? Describe three relay applications in transmission systems.

10. What is meant by the BIL of a piece of equipment?

5

Extra-High Voltage and Direct-Current Transmission

NEED

Increase in transmission voltages are due to many factors, but principally because of the need to provide ever greater amounts of power over the same rights-of-way. As demands for power increase, as the cost of land and its clearing increase and its availability decreases, and as power systems continue to be interconnected to provide greater economy and service reliability, so too does that need become more and more imperative.

PROBLEMS

While such increase in the voltage of a transmission line appears simple, it creates problems involving insulation, lightning protection, switching, and losses.

Equipment, such as circuit breakers, transformers, surge arresters, reactors and capacitors, towers and supports, conductors, and almost every other item of material, are all affected. Corona losses, skin effect, effect of moisture, dirt, smog and salt on insulators, all must be considered.

Audible noise, as well as radio, television, and other communications, also tend to increase with higher overhead higher voltage transmission lines.

Questions also are raised as to the effect of such high-voltage lines on human, animal, and plant life, as well as on structures, vehicles, and other organic and nonorganic objects located or operating in their vicinity.

Undergrounding high-voltage transmission lines proves technically feasible but economically unfeasible, except in a few instances such as for very short distances within such cities as New York.

Direct-current transmission for long overhead or short underground or underwater distances is generally limited to intersystem ties of from 250 to 1500 megawatt capacities.

OPERATIONS

There is the question of limits to the capacity of transmission lines so that its outage or nonavailability will not inordinately affect the operation of the system or grid of which it is a part. This compares to the effect of a large size generator out of service.

Further, for overhead lines which constitute the overwhelmingly majority of such lines, the maximum achievable transmission line loading is limited by the right-of-way available.

It is evident, therefore, that the national environmental concerns, including beautification, and the desire for low-cost power are not easily reconciled.

Relationship of Line Capability and Voltage

For a given conductor, that is, a conductor with a fixed electrical resistance, the amount of electricity or current that may be carried by it economically is limited by either (1) the permissible *voltage drop* or loss of electric pressure or (2) the *power loss* which represents the heat dissipated from the line. Both must be considered in relation to either the transmission line voltage or total power transmitted. For example, permissible voltage drop may be 10 percent of the transmission line voltage, or permissible power loss may be 10 percent of the total power transmitted.

$$\text{transmitted power} = \text{current} \times \text{voltage}$$

or

$$P = I \times E_{\text{line}}$$

Then, if the voltage of a line is doubled, the current that will flow for the same power supply will be only half the original current.

Voltage drop and power loss may be expressed as follows; where R is the resistance of the conductor, E is the voltage drop or voltage loss, P is the power loss in the transmission line:

$$E_{\text{drop}} = IR \qquad \text{and} \qquad P_{\text{loss}} = I^2R$$

With half the current flow, the voltage drop or loss is also halved, and the power loss, with only half the current flow, will be only one quarter the original loss. Conversely, if the power loss can be maintained at its original proportional part, the transmitted power can be quadrupled. The voltage then can be doubled, and the amount of current sent through the same conductor can also be doubled. The power which this conductor now is capable of carrying, represented algebraically, as compared to the original is:

$$P(\text{orig.}) = I(\text{orig.}) \times E(\text{orig.})$$

$$P(\text{new}) = 2I(\text{orig.}) \times 2E(\text{orig.}) = 4P(\text{orig.})$$

Hence, doubling the voltage of a line will increase its capability four times.

Extra-High Voltage

Many economic factors have contributed to the development of higher and higher voltage transmission lines. Basically, the power that a line will carry increases as the square of its voltage. If the voltage is doubled, the power capacity is increased four times. More specifically, a 345 kV line will carry nine times the power of a 115 kV line even though the voltage is only three times as high. A 765 kV line has 25 times the power capability of a 138 kV line. When all the factors are considered, the 765 kV is capable of carrying four to six times as much power as one at 345 kV over comparable distances. Put another way, the cost per megawatt of power transmitted is reduced considerably when a higher voltage is used. This is illustrated in Figure 1.5 in Chap. 1. Tower lines and equipment are very large in size (Fig. 5.1).

COST COMPARISONS

In selecting the voltage of a transmission line, it is essential that the incremental differences in costs of all the other associated equipment be taken into consideration; e.g., circuit breakers, transformers, switchgear, lightning protection, structures, buildings, rights-of-way, the various control equipment and measuring devices, etc. Hence, annual carrying charges including annual losses of energy from all sources, as influenced by the different voltages under consideration, must be taken into account. Results should, however, be tempered with practical considerations, such as availability of skilled workers, access problems, and, most especially, in protecting the environment, appearance, and the effect on the community.

  Comparing costs with alternating-current systems of similar voltages and load carrying abilities, while costs for terminals and other associated equipment are essentially the same, line costs can be substantially lower (Table 5.1).

Figure 5.1 765KV guyed V-aluminum tower line (170 ft high). Each phase conductor consists of a bundle of four substrands. (*Courtesy of Ohio Brass Company, a subsidiary of Henry Hubbell.*)

TABLE 5.1 Economic Comparisons for Extra-High Voltage Transmission (in percent)

Equipment for AC Transmission	450 kV	765 kV	1000 kV
Circuit breakers/unit	100	200	400
Autotransformers/kVA	100	125	150
Reactors/kVAR	100	120	140
Series Capacitors/kVAR	100	110	120
Line/mile	100	150	210
Equipment for DC Transmission		± 375 kV	± 900 kV
Terminals, including rectifiers/kW		200	220
Line/mile*		50	200

*Compared to AC figures

Mine-mouth Generation

The use of extra-high voltages makes practical the so-called mine-mouth generation (Fig. 5.2). Here, instead of transporting coal to the generating plant, it is cheaper to generate the power right at the mine and then transport the power by transmission line. Similarly, other remote sources of power, as hydroelectric plants, are made practically available to major consumer centers. Because of insulating material limitations, voltages produced by generators are restricted to about 20 kV. The alternating-current (ac) output, however, is readily changed by means of transformers, so that the generated voltages are easily stepped up to transmission voltages.

EQUIPMENT

Circuit Breakers

Circuit breakers are becoming more complex. Because the higher voltages tend to have longer arcs and persist for longer periods of time during the disconnecting phase of their operation, greater attention is paid to the insulating arc-quenching medium. Use of sulphur hexafluoride (SF_6) gas under pressure has solved the problem of arcing, at the same time providing insulation qualities that are some hundred times better than oil [see Fig. 3.5(c), Chap. 3]. The pressure applied to the gas, however, aggravates the possibility of leakage.

The sudden interruption of current at these high voltages creates voltage surges greater in magnitude than line voltages. These surges not only are applied to the breaker conductors, but travel out along the lines until they dissipate or are bled to ground through surge arresters. The effect of this phenomenon is to require greater insulation to be specified at critical points in the breaker, the lines, and in other equipment that may be connected to them. The interrupting duty of such breakers becomes greater as values of voltage and current become larger, requiring much greater mechanical reinforcement of the component parts. The size of the breaker may be inordinately large.

Transformers, Reactors, Capacitors

Similar problems of higher voltages, surges, and mechanical stresses imposed on component parts apply to these pieces of equipment as to circuit breakers (detailed in previous section). Structural reinforcement of coils, cores, and supports are necessary to withstand the extraordinarily high currents that may flow when faults or short circuits occur anywhere on the line.

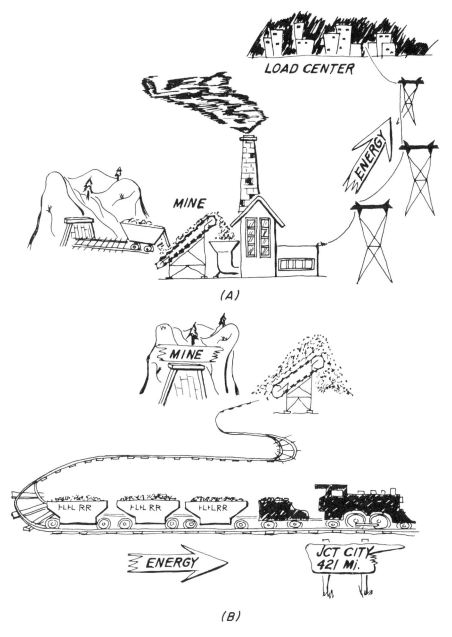

Figure 5.2 (a) In mine-mouth power generation, power is generated at the mine and then transported; (b) rather than transporting coal to generate energy elsewhere.

As no arcs are to be quenched, oil or askarels are satisfactory for insulation purposes. Because of possible voltage surges causing flashover, bushings may be proportionately larger than those for lower voltage equipment.

Lightning or Surge Protection

Voltage surges caused by lightning or switching are more apt to result in flashover across insulator strings and to the tower or support structure, and involve problems of clearances and insulator swing geometrics.

Lightning striking at or near an overhead transmission line creates surges of voltage that travel along the line. This action is the same for all overhead lines exposed to lightning. Extra-high-voltage transmission lines differ from lower voltage lines in that the line current that flows in the flashover that may occur across insulators or insulation at some point tends to be much greater because of the greater line voltage. Obviously, this greater current imposes greater stresses on the equipment through which it may flow, with possible damage to the equipment. This is especially true of the circuit breakers called upon to interrupt such currents.

Not only are a greater number of lightning or surge arresters installed along the line, but greater effort is made to reduce ground resistances to enable the surge voltage to be dissipated as quickly as possible.

While the voltage surges caused by switching may not be as great as those caused by lightning, the same general conditions result as those described, and the solutions are the same.

To aid in reducing damaging flashovers, extra insulators may be added to the strings than would normally be required. The longer string of insulators allows the conductor attached to it to have a greater radius in its sway. This not only calls for greater clearances of the conductor from the tower or supporting structure, but the string must be carefully placed so that the swinging conductor will not make contact with adjacent conductors.

For the principle of operation of surge arresters, the several types, and their application, refer to Chap. 2, Overhead Construction.

UNDERGROUND

As mentioned earlier in Chap. 4, the capacitance or condenser effect of high-voltage cable absorbs energy and limits the amount of useful energy as well as the distance over which it may be transmitted. To reduce this capacitance effect, compressed gas (sulphur hexafluoride) at 50 pounds per square inch pressure is employed as insulation in extra-high-voltage cables, and while satisfactory for this purpose, it introduces problems of maintenance (leakage and repairs). Nevertheless, such form of insulation can withstand the great stresses of such high voltages under varying temperature and current loading conditions.

Super Conductors

Liquified gas such as nitrogen circulating within the conductors to maintain extremely low temperatures (cryogenic) make possible conductors having very low resistance. For example, if aluminum is cooled to the temperature of liquid nitrogen ($-320°F$), its electrical resistance is approximately one-tenth the value at normal temperatures; its mechanical properties change at such low temperatures. However, necessary refrigeration and associated equipment to maintain the liquidity of the gas for the continuous removal of heat requires special manufacture and use of certain materials, resulting in an extremely costly installation.

Certain comparatively rare metals, such as Niobium, if cooled to temperatures approaching that of liquid helium ($-425°F$), lose all resistance to the flow of direct-current. Such cables are termed super conductors (as compared to the cryogenic cable mentioned), the difference being in the temperatures and materials employed.

Economics indicates that such cables can become practical, offering the possibility of substantial reduction in cost in the high power capability range from about 5000 mV·A and above.

DIRECT-CURRENT TRANSMISSION

Comparison of AC and DC

In a circuit, the maximum voltage permissible is fixed by its insulation. In dc systems, the maximum voltage is applied throughout the entire time the flow of electricity takes place. In ac systems, the maximum voltage is applied only a portion of the time. If the average was taken of all of the values, from zero to maximum to zero, it would be found that this average value is 70.7 percent of the peak or maximum value (Fig. 5.3). Hence, the capacity of a two-wire ac system is limited to only 70.7 percent of that of a two-wire dc system having the same insulation, and the same conductor size, in order that the insulating value of the insulation not be passed.

DC Transmission Features

Many benefits can be realized by using direct-current (dc) for high-voltage transmission. Where, in ac circuits, the effective voltage is 70.7 percent of the peak value the line carries, in dc circuits, the effective and peak value are

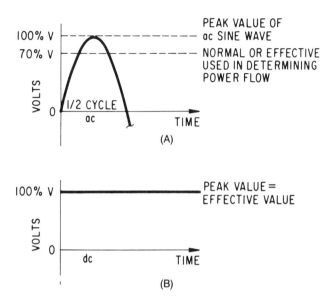

Figure 5.3 Ac versus dc power transmission (top). In ac, the peak voltage must be used in calculating the insulation required from conductor to grounded supporting structure. This value is higher than the effective value (bottom). The dc system utilizes the maximum voltage to ground to transmit power.

one and the same. Hence, for a particular voltage rating, the dc circuit requires only 70.7 percent of the insulation required by the ac circuit; conversely, with the same size cables and the same insulators, a dc line can carry some 40 percent more power. Further, the conductors are not subject to skin effect, although corona continues to be a problem. Since the circuits are not subject to alternating magnetic fields, no inductance within a conductor or between conductors is generated (except at the moments of energization and deenergization); hence, voltage and energy losses are reduced. Further, as will be described later, fewer conductors are employed in a dc circuit as compared to an ac circuit. Generally two or three are used in direct-current as compared to four in alternating-current. In some cases, where "ground" is used as a return circuit (such as a body of water), as few as only one dc conductor may be used.

Advantages

High-voltage direct-current permits the use of higher effective voltages for a given number of insulators in a string and a given length of flashover distance between the conductors and supporting structures. Leakage and corona losses are thereby decreased.

Since there is no alternating magnetic field around the conductor (except at the time of energizing and deenergizing) there are no inductive or capacitive effects that, in alternating-current systems, often require corrective measures. The power factor is unity, a fact that alone accounts for considerable reduction of transmission losses. This is especially important in underground cable systems. Similarly, the lack of a moving magnetic field reduces the stresses produced on circuit breakers, reducing their interrupting duty requirements.

Another important advantage of direct-current transmission is the ability to interconnect separate transmission systems (of the same voltage) without the necessity of first synchronizing them (Fig. 5.4). This essentially eliminates the stability problems associated with interconnected alternating-current transmission systems described in Chap. 4, Figure 4.22(a),(b).

Since such lines may consist of fewer conductors (use of only two conductors, or even one compared to three or more for alternating-current systems), tower and supporting structure requirements as well as those for rights-of-way are relatively less and reflected in lower line construction and maintenance costs.

Disadvantages

It is generally uneconomical to tap direct-current lines and tapping is generally avoided. Line tapping requires conversion to alternating-current before lower (or higher) voltages can be transformed to desired values. If the desired voltage is direct current, conversion from alternating-current back to direct-current is also required [Fig. 5.5(a),(b)].

Changing Voltage—Direct-Current Systems

To raise or lower voltages in a direct-current system, it is necessary to go through the same procedure described for tapping of such circuits. Costly terminal equipment is necessary to convert power to alternating-current at transmission and distribution substations to make direct-current lines part of

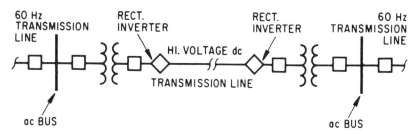

Figure 5.4 Direct-current transmission used over very long distances. The two ac buses may be hundreds of miles apart, and do not have to be synchronized, or in phase, to permit power flow between systems.

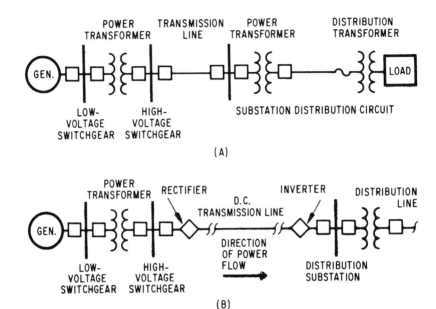

Figure 5.5 (a) Arrangements of ac-power system links; and (b) dc-power system links.

the overall electric system which ultimately supplies commercial distribution. Further, whether tapping direct-current lines or changing values of direct-current voltages, it has not yet proven feasible to control power flow and protect the circuit in the same fashion as a tap-off or change in voltage of an alternating-current circuit.

Rectifiers

Conversion of alternating-current power to direct-current (and vice versa) is accomplished through rectifiers. For the amounts of power to be transmitted in such extra-high-voltage lines, mercury arc rectifiers, sometimes referred to as mercury valves, are used, with the number employed depending on the amount of power to be converted. Such mercury arc rectifier installations are usually large and require periodic maintenance. Moreover, the stability of the arc is subject to momentary dips in the incoming source which may cause the arc to be extinguished, necessitating their restarting. Later installations employ solid state-type rectifiers, sometimes referred to as thyristor valves, replacing the mercury arc converters. While these, too, are costly, they require little or no maintenance.

Summing, high-voltage direct-current is suitable almost exclusively for transmission of large blocks of power over long distances in point to point transmission. While the cost for direct-current lines is less than for alternating-current, the cost of the converter stations necessary for the conversion of alternating-current to direct-current for transmission and then back to alternating-current, brings the total cost to a value more than that for alternating-current transmission for short lines. Economically, a break-even point in costs for high-voltage direct-current compared to extra-high-voltage alternating-current appears to be at line lengths of about 400 to 500 miles, at which point direct-current appears economically feasible. As for underground cable circuits, this break-even point appears to be some 40 to 60 miles.

History

The earliest transmission lines were short, low voltage, low-power direct-current runs, confined almost solely to European countries, consisted of a series circuit, known as the Thury system (Fig. 5.6). Here a number of direct-

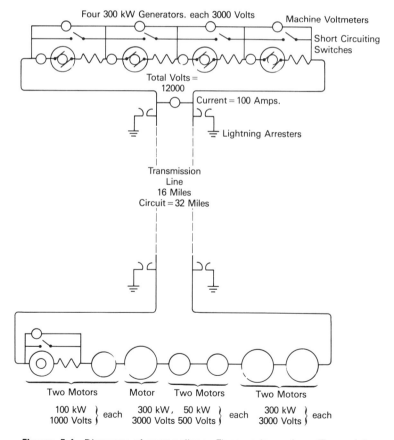

Figure 5.6 Diagram of connections—Thury series system. (Example)

current generators were connected in series to develop the high voltage of the transmission line, and carrying a constant value of current. At the receiving end, a number of motors connected in series, located at one or more substations, with the current flowing in each one having the same value. These motors, in turn, drove direct-current generators whose output served low-voltage direct-current distribution requirements. Variations in load, which necessitated changes in the voltage to vary with the load, were taken care of by short circuiting one or more generators as well as motors at the substations. This system is shown diagrammatically and as an example in Figure 5.6.

The vast increases in demands for electricity, mostly because of World War I, largely were more readily met by the rapidly developing and expanding conventional alternating-current systems, employing the transformer (with no moving parts). Thury systems were finally supplanted by conventional alternating-current systems, although the last ones persisted until the early 1930s.

REVIEW

☐ Extra-high-voltage (EHV) transmission often permits transmitting large blocks of power more economically than lower voltage transmission.

☐ Effective voltage of an alternating-current is only 70.7 percent of the maximum voltage of the alternating-current sine wave.

☐ EHV transmission is ideal for "mine-mouth" power generation, whereby the power-generating station is located right at a coal mine. This eliminates having to transport the coal to a generating station miles away (Fig. 5.2).

☐ Load carrying ability of transmission lines varies as the square of the operating voltages; doubling the line voltage results in quadrupling the line capacity.

☐ EHV transmission lines have serious effects on the design and construction of related equipment, such as circuit breakers, transformers, reactors, capacitors, lightning or surge arresters, conductors, towers and supporting structures, etc.

☐ EHV lines are subject to insulator flashovers and subject equipment to large mechanical stresses.

☐ Flashovers caused by lightning or switching may result in greater flow of "follow" line current that may cause greater fault current to flow, imposing greater stresses in equipment.

☐ Liquified gases at extremely low temperatures (in the −400°F range) circulating within conductors causing their resistances to approach zero in value; these are often referred to as cryogenic cables and super conductors.

☐ The major advantages of direct-current transmission over alternating-current transmission are: (1) higher power capability for the same size conductor and insulation; (2) elimination of skin effect and inductances, loads are delivered at essentially unity power factor; and (3) fewer conductors required for transmission.

☐ The major disadvantage of direct-current transmission is the difficulty in stepping voltages up and down, requiring expensive equipment, including rectifiers and associated equipment.

STUDY QUESTIONS

1. What is meant by extra-high-voltage transmission?
2. What is the relationship between line voltage and capability?
3. What three main considerations are taken into account in selecting the type and voltage of a transmission line?
4. What are the advantages of EHV transmission?
5. What are the disadvantages of EHV transmission?
6. What is meant by "mine-mouth" generation? Where is it used?
7. What are the advantages of direct-current transmission compared to alternating-current transmission?
8. What are some of its disadvantages?
9. What are super conductors? What are their advantages? What are some of their disadvantages?
10. How may alternating-current and direct-current systems be interconnected?

6

Basic Electricity

BASIC ELECTRIC CIRCUIT

The basic circuit for transmitting electrical energy must consist of two paths or conductors, one sending and the other a return, together forming a continuous path or a closed circuit for the electric current to flow. This holds for both direct- and alternating-current systems.

The conductors or wires in an electric circuit can be thought of as electrical pipes through which flow a stream of electrons constituting a flow of electricity (refer to Chap. 1 for a water analogy).

Electric Pressure or Voltage

For the electrons to stream through the wires, an electrical pressure is necessary. This pressure can be created by a generator which may be likened to a pump. The electrical pressure is expressed in volts (named in honor of the Italian physicist Alessandro Volta).

Current or Amperage

Electric current is the rate of flow of electrons in a circuit. Similar to the flow of water measured in gallons or litres per second, the number of electrons passing a reference point in one second determines the current strength, and is expressed in amperes (named after the French physicist Andre Marie Ampere). Scientific measurement has determined that 6.29 billion billion electrons passing in one second make up one ampere.

Resistance

Just as water flowing in a pipe is resisted by friction between the water and the pipe, electrons encounter resistance to their flow in conductors. As a fixed volume of water encounters less resistance to its flow in a large diameter pipe than in one of small diameter, so, too, a fixed value of electric current will encounter less resistance in a large diameter conductor than in one of small diameter. The electrical resistance is expressed in ohms (after George Simon Ohm, the German philosopher).

OHM'S LAW

In any electrical circuit, the three factors of pressure, current, and resistance are interrelated. The relationship may be expressed as follows:

In a circuit, the current flow will vary directly as the pressure applied, and inversely as the resistance of the circuit. Expressed in the terms defined

$$\text{Current in amperes} \ = \ \frac{\text{Pressure in volts}}{\text{Resistance in ohms}}$$

This is known as Ohm's Law. Expressed as an equation, the law becomes

$$I \ = \ \frac{E}{R}$$

where I is the intensity of the current in amperes, E the pressure in volts, and R the resistance in ohms. If any two quantities are known, the third may be found by applying the equation above.

TYPES OF CIRCUITS

There are two basic types of electric circuits: the series circuit and the multiple or parallel circuit; other types are a combination of the two.

Series Circuit

In a series circuit, all the parts making up the circuit are connected in succession, so that the current through all of the parts is the same.

The circuit shown in Figure 6.1 contains four resistances connected in

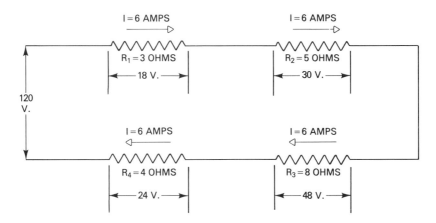

Figure 6.1 Simple series circuit.

series with the same current through each of them. Assume an electrical pressure or voltage of 120 volts is applied across the terminals of the circuit and a current of 6 amperes flows through the circuit. Then, by Ohm's Law, the total resistance of the circuit will be found to be 20 ohms; that is,

$$I = \frac{E}{R} \quad \text{or} \quad 6 = \frac{120}{R} \text{ ; solving for } R = \frac{120}{6} = 20 \text{ ohms}$$

Voltages measured across each of the four resistances, R_1, R_2, R_3, and R_4, shown are found to be

$$E_1 = 18 \text{ volts; } E_2 = 30 \text{ volts; } E_3 = 48 \text{ volts; } E_4 = 24 \text{ volts}$$

Applying Ohm's Law to each part of the circuit, the resistance values can be found

$$I = \frac{E}{R} \quad \text{from which} \quad R = \frac{E}{I}$$

Hence

$$R_1 = \frac{18}{6} = 3 \text{ ohms} \qquad R_2 = \frac{30}{6} = 5 \text{ ohms}$$

$$R_3 = \frac{48}{6} = 8 \text{ ohms} \qquad R_4 = \frac{24}{6} = 4 \text{ ohms}$$

Checking: the sum of the four voltages across each of the resistances (the drop in pressure as the current flows in the resistance) will be:

$$E_1 + E_2 + E_3 + E_4 = E$$

or

$$18 \text{ volts} + 30 \text{ volts} + 48 \text{ volts} + 24 \text{ volts} = 120 \text{ volts}$$

And the sum of the separate resistances is equal to the total resistance.

$$R_1 + R_2 + R_3 + R_4 = R$$

or

$$3 \text{ ohms} + 5 \text{ ohms} + 8 \text{ ohms} + 4 \text{ ohms} = 20 \text{ ohms}$$

Note that there is a drop in voltage through each resistance. From Ohm's Law, this drop will be the product of the current and resistance, that is $E = IR$, and is usually referred to as the *IR* drop. Note also that the voltage drop in each part is proportional to its resistance and that the sum of all the voltage drops is equal to the applied voltage. (Resistance of the connecting wires has been neglected for purposes of illustration.)

Multiple or Parallel Circuits

In a multiple or parallel circuit, all of the components are connected so as to receive full line voltage, and the current that flows in each component depends on its resistance.

If the same four resistances are connected in parallel across the same line voltage, as shown in Figure 6.2, applying Ohm's Law, the currents flowing in each resistance have the following values:
Since

$$R_1 = 3 \text{ ohms}, I_1 = \frac{E}{R_1} = \frac{120}{3} = 40 \text{ amperes}$$

$$R_2 = 5 \text{ ohms}, I_2 = \frac{E}{R_2} = \frac{120}{5} = 24 \text{ amperes}$$

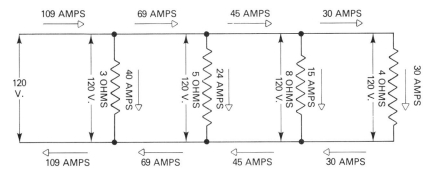

Figure 6.2 Resistance in multiple circuits.

$$R_3 = 8 \text{ ohms}, \quad I_3 = \frac{E}{R_3} = \frac{120}{8} = 15 \text{ amperes}$$

$$R_4 = 4 \text{ ohms}, \quad I_4 = \frac{E}{R_4} = \frac{120}{4} = 30 \text{ amperes}$$

The total current will be:

$$I_1 + I_2 + I_3 + I_4 = I$$

or

$$40 \text{ amperes} + 24 \text{ amperes} + 15 \text{ amperes} + 30 \text{ amperes} = 109 \text{ amperes}$$

The resistance of the entire circuit, applying Ohm's Law, will be

$$R = \frac{E}{I} = \frac{120}{109} = 1.113 \text{ ohms}$$

Another way of obtaining the same result is to add the reciprocals of each resistance and taking the reciprocal of the sum of the reciprocals.

$$\frac{1}{R_1} + \frac{1}{R_2} + \frac{1}{R_3} + \frac{1}{R_4} = \frac{1}{R}$$

$$\frac{1}{3} + \frac{1}{5} + \frac{1}{8} + \frac{1}{4} = \frac{1}{R}$$

$$0.333 + 0.200 + 0.125 + 0.250 = 0.908$$

$$R = \frac{1}{0.908} = 1.113 \text{ ohms}$$

Note that the resultant resistance is always less than the smallest of the component resistances. Each additional resistance connected in parallel adds an additional path for the current to flow; as the conducting paths are increased, the total resistance is lowered.

Series-Parallel Circuit

An example of resistances connected in series–parallel is shown in Figure 6.3. To obtain the total resistance of this circuit, the resultant resistance of each of the parallel groups is first determined, then added to the resistances in series. The same process applies to any number and kind of groups of resistances.

EXERCISE: Find the voltages across each of the resistances, the currents flowing in them, and the total resistance of the circuit.

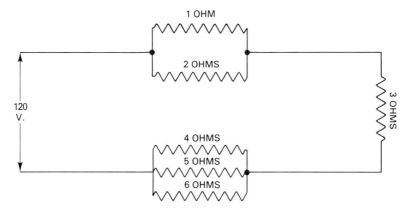

Figure 6.3 Series multiple circuits.

POWER

When electricity flows through a resistance, electrical energy is consumed in that resistance. The *rate* of consumption of electrical energy is known as electrical power. The unit of electrical power is called the watt (W) (after James Watt, the developer of the steam engine); the watt is too small for practical purposes and the kilowatt (kW) equal to 1000 watts is most frequently used.

Power is the *rate* of doing work. Electrically, this depends on the electrical pressure applied and the current flowing in a device, represented as a resistance. Hence,

$$\text{Power (watts)} = \text{Pressure (volts)} \times \text{Current (amperes)}$$

By experiment, it is found that 746 watts of electrical power is equivalent to one horsepower of mechanical power.

ENERGY

Since power is the rate of expending energy, then energy expended will be the product of power and the time it is applied

$$\text{Energy} = \text{Power} \times \text{Time}$$

Time may be expressed in any unit: seconds, minutes, hours, days, etc. The common unit is the hour, hence

$$\text{Energy} = \text{Power (kilowatts)} \times \text{Time (hours)}$$

The most commonly used unit of electrical energy is the kilowatt-hour (kW·h).

HEAT LOSS

When power flowing in a circuit does not produce useful work, that is, where the electrical energy is not converted to some mechanical work, it is converted into heat and dissipated into the surrounding atmosphere. This heat, owing to the electrical resistance encountered, may be likened to the heat developed by friction, and represents a loss. This may be determined by applying Ohm's Law:

If

$$\text{Power in watts} = \text{Voltage} \times \text{Current}, \quad \text{or} \quad W = E \times I$$

and

$$\text{Voltage} = \text{Current} \times \text{Resistance}, \quad \text{or} \quad E = IR$$

then

$$\text{Power (W)} = \text{Voltage } (IR) \times \text{Current } (I)$$
$$W = IR \times I$$
$$W = I^2R$$

This represents the rate of energy loss given off as heat; actual energy loss multiplies this value by the time it is expended and is measured in kilowatt-hours.

Note that if the current (I) flowing in a resistance (conductor) is doubled, the heat loss is not doubled, but quadrupled, or four times as much.

In transmitting power over great distances, two factors must be considered.

1. The voltage (IR) drop in the line must not be so great that insufficient electrical pressure or voltage will result at the receiving end.
2. The power (I^2R) loss in the line must not be so great that other means of supplying power may prove more economical (e.g., set up a generating plant at or near the receiving end and transport fuel there).

INDUCTANCE

There is a basic relationship between electricity and magnetism. An electric current flowing in a conductor will produce a magnetic field around it. A conductor cutting the magnetic field will have an electric voltage induced in it and a current will flow if the conductor is part of a circuit. In the latter case, the magnitude of the voltage produced depends on the length of the conductor cutting the field, the speed at which the conductor cuts the magnetic field, and the density of the magnetic field.

A conductor carrying an alternating current will have a magnetic field around it that alternates its characteristics as the current in the conductor alternates with its frequency. The magnetic field builds up to a maximum in one direction, reduces to zero, builds up to a maximum in the opposite direction and again reduces to zero, completing one cycle. (Frequency is the number of such alternations, or cycles, occurring in one second.) (See Fig. 6.4.)

Self-Inductance

The alternating magnetic field around the conductor will cut the conductor, inducing in it a voltage distinct from that causing the original current to flow. This second voltage will also produce a current which, in turn, affects the original current. In turn, this affects the magnetic field around the conductor, thus affecting the entire relationship which finally stabilizes at some point.

The current flowing in the conductor, the original and that induced, are not in step, or "in phase"; that is, the rising and falling of their cycles do not coincide. Their relationship may be shown in Figure 6.5; the two voltage waves will be displaced a quarter cycle from each other. Actually, of course, only the one resultant voltage exists in the conductor.

The current values, however, still reach their maximum and zero values as they did originally. The net effect of the reaction in the conductor is to cause the current to "lag" behind the voltage, as illustrated in Figure 6.5(c). It will be noted then that the current and voltages do not act together throughout the cycle, but only some portion of the current will act in conjunction with the voltage. The power produced then will be the product of the voltage and current values at any particular point and not the product of the voltage and current when acting together, or "in phase." The ratio of the first quantity to the latter is known as the "power factor."

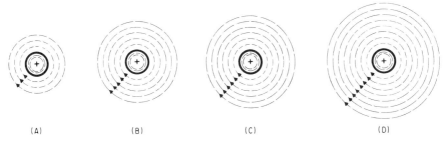

(A)   (B)   (C)   (D)

Figure 6.4 a, b, c, d—Magnetic field expanding about a conductor. d, c, b, a—Magnetic field contracting about a conductor.

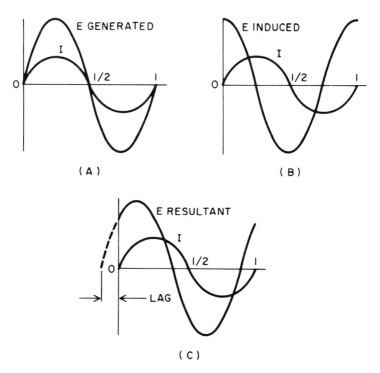

Figure 6.5 Effect of inductance on voltage and current in a conductor (not to scale).

Mutual Inductance

This same reaction may be caused by the magnetic fields of adjacent conductors and, since both conductors affect each other, it is called "mutual inductance." (See Fig. 6.6.)

Inductive Reactance

The effect of this inductance is to act as an obstruction to the flow of current in a circuit, similar to but greatly different than the effect of resistance. Such an obstruction is known as the "reactance" of the circuit, and is also expressed in ohms. (To avoid confusion, the letter R is used to denote resistance and X_L to denote reactance due to inductance.)

Ohm's Law can also be used to find the current flow in an inductive circuit, assuming the circuit has no resistance. Hence

$$I = \frac{E}{X_L}$$

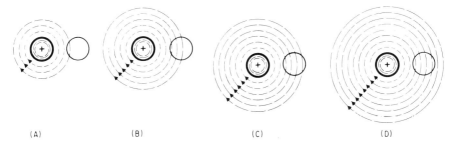

Figure 6.6 Illustrating the effect of the magnetic field about a conductor on an adjacent conductor.

Note, however, that when only inductance obstructs the flow of electricity, the current wave "lags" the voltage wave by a quarter cycle [Figure 6.5(b)].

Resistance and Inductance

Applying Ohm's Law to each of two circuits, one containing resistance only, the other inductive reactance only: assume a voltage of 120 volts applied to each presenting an obstruction equivalent to 2 ohms, then

$$I_1 = \frac{E}{R} = \frac{120}{2} = 60 \text{ amperes} \qquad I_2 = \frac{E}{X_L} = \frac{120}{2} = 60 \text{ amperes}$$

Now assume a circuit that has a combination of both resistance and inductive reactance, with the same values indicated above. With a voltage of 120 volts applied, the current will be found to be 42.6 amperes. By Ohm's Law,

$$R = \frac{120}{42.6} = 2.82 \text{ ohms}$$

The resultant obstruction to the flow of electricity is greater than either the resistance or the inductive reactance (each 2 ohms) but is *not* equal to the arithmetic sum (4 ohms). By analysis, it will be found that the resultant obstruction may be found by obtaining the square root of the sum of the squared values of resistance and inductive reactance, that is,

$$Z = \sqrt{R^2 + X_L^2} \quad \text{or} \quad \sqrt{2^2 + 2^2} = \sqrt{8} = 2.82 \text{ ohms}$$

where Z is the resultant obstruction and is referred to as "impedance" to differentiate it from its component resistance and inductive reactance.

From Figure 6.7b it will be noted that when resistance is combined with inductance, the current lags the voltage wave less than when the circuit contained only inductance. As more resistance is added, the current and voltage waves will more closely approach each other.

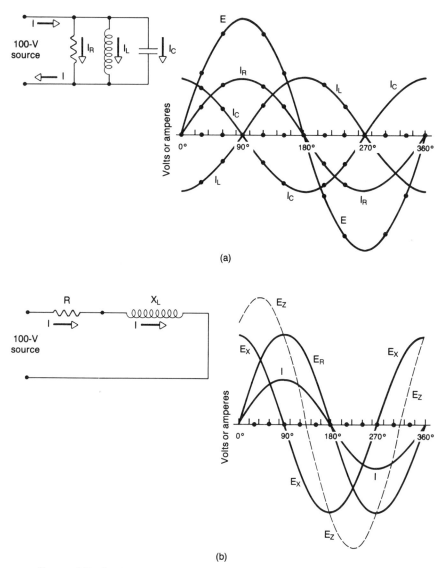

Figure 6.7 Cycle of voltage and current values: (a) resistance, inductance, and capacitance in parallel; (b) resistance and inductance in series.

CAPACITANCE

There is another reaction that occurs in conductors carrying an alternating-current that is *not* due to the magnetic fields around them. This is an electrostatic effect that takes place between conductors. During one-half cycle of the alternation, there will be a scarcity of electrons in a conductor (with reference to a fixed point) and it will tend to attract electrons from an adjacent conductor. During the next half cycle, a reverse action takes place—an excess of electrons in the conductor, and there will be a tendency for electrons to flow to the adjacent conductor. Thus a to-and-fro, or alternating, circulation of electrons is set up in the second conductor; that is, an alternating current is set up in the second conductor.

If both conductors carry alternating-current, they will react on each other as described above. The amount of this reaction is called "capacitance" and will depend on the areas of the conductors exposed to each other. The distance and kind of insulating material between them, and the number of electrons involved depend on the voltage or current in the conductor.

Similar to inductance, capacitance will set up a distinct current in the conductor which will be displaced by a quarter cycle from the normal line current. In this case, however, the current will "lead" the voltage wave by a quarter cycle (Fig. 6.8).

Such obstruction of the capacitance or capacitor to the flow of electricity in a conductor is referred to as capacitive reactance and is also measured in

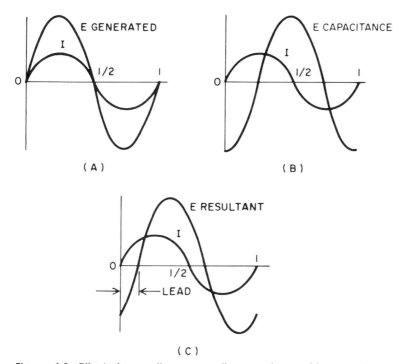

Figure 6.8 Effect of capacitance on voltage and current in a conductor (not to scale).

ohms. Capacitive reactance is denoted by the letters X_C to distinguish it from other quantities.

Water Analogy of a Capacitor

Refer to Figure 6.9. As the pump moves to the right, water flows in the top pipe to the cylinder at the right. The flexible diaphragm in the cylinder will move to the right (as shown), transmitting its motion to the water to its right during the motion of the pump. In the leftward motion of the pump, the reverse action takes place. The net result is that an alternating-current flow is set up in both the pipes, without a direct connection between the water in the lower pipe and that in the upper pipe.

In the process, because of the flexibility of the diaphragm, the current of water will tend to arrive at the undistended diaphragm before the pump completes its stroke in one direction. This may be observed by the relative positions of the driving motor on the left and the water in the driven machine on the right. Compared to the flow of water in the pipes driving a water motor (see Figure 6.16) the current flow may be said to "lead," or act ahead of, the pressure being applied by the pump at the left. The mechanical stresses on the diaphragm, alternately compression on one side and elongation on the other, correspond to the electrostatic stresses on the dielectric in the capacitor.

Resistance and Capacitance

The same phenomenon as existed in an inductive circuit exists in the capacitive circuit, except that the obstructive effects are in opposition or 180° apart. In a circuit containing both resistance and capacitance, the resulting obstruction is obtained in the same manner as for resistance and inductance, that is,

$$Z = \sqrt{R^2 + X_C^2}$$

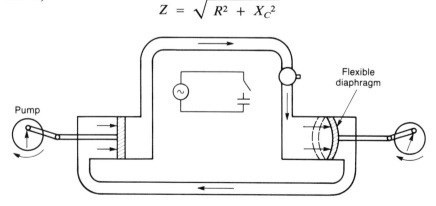

Figure 6.9 Water analogy of a capacitor in an alternating-current circuit.

In this case, the more resistance added, the closer will the voltage and current waves approach each other.

IMPEDANCE

The total obstruction to the flow of current in a circuit may be caused by resistance, inductance, and capacitance. The effects of inductance and capacitance are generally referred to as reactance. Hence, the combined effect of resistance and reactance is called the "impedance" of the circuit and, as already noted, is designated by the letter Z. For alternating-current, therefore, Ohm's Law becomes

$$I = \frac{E}{Z}$$

RESISTANCE, INDUCTANCE, AND CAPACITANCE

As mentioned earlier, the effects of inductance and capacitance are in direct opposition to each other. Hence, the net reactance will be the difference between the inductive reactance and capacitive reactance, that is,

$$X = X_L - X_C$$

The resultant impedance of a circuit containing resistance and reactance will then be

$$Z = \sqrt{R^2 + X^2}$$

or, substituting

$$Z = \sqrt{R^2 + (X_L - X_C)^2}$$

and the current flowing in the circuit will be by Ohm's Law

$$I = \frac{E}{Z} = \frac{E}{\sqrt{R^2 + (X_L - X_C)^2}}$$

Resonance

It is evident that if $X_L = X_C$, the only quantity left will be R, the resistance. When this condition occurs, the circuit is said to be in "resonance."

The relative value of resistance, inductance, and capacitance of a circuit will determine the relative position of the current wave with respect to the voltage wave.

POWER IN ALTERNATING-CURRENT CIRCUITS

As mentioned earlier, power is the rate at which electrical energy is transformed into heat or mechanical energy and is equal to the product of the voltage and current. In an alternating-current circuit, the change of energy at any moment is the product of voltage and current at that instant. Power values for circuits having only resistance, only inductance, and only capacitance, are shown in Figure 6.10. Note that real power is produced only in the resistance circuit. In the inductance and capacitance circuits the net result of the power curves is zero, that is, the areas under the curves above the axis are exactly equal to those below the axis.

In both the inductance and capacitance circuits, electric energy is stored in the magnetic and electrostatic fields during the time the current is increasing and returned back to the circuits when the current is decreasing.

POWER FACTOR

As indicated earlier, the product of voltage and current does *not* indicate the true power flow in the circuit. Such a product is termed "apparent power" and is expressed in volt-amperes, or kilovolt-amperes (kVA) instead of watts, or kilowatts (kW). To calculate the true or real power, the apparent power is multiplied by a factor called the "power factor."

The power factor of a circuit is the ratio between the true and the apparent power and is usually expressed as a decimal or percent.

$$\text{Power Factor} = \frac{\text{True power}}{\text{Apparent power}} = \frac{\text{Watts}}{\text{Volt-amperes}} = \frac{\text{kW}}{\text{kVA}}$$

In the resistance circuit, all of the energy delivered to it is converted into heat or mechanical energy; hence the true and apparent power are the same and the power factor is 1, unity, or 100 percent. In the inductance or capacitance circuit, the true power is zero while the apparent power may have a value; the power factor for both, however, is 0, zero, or 0 percent. In a circuit containing resistance, and inductance and capacitance, the real power and apparent power are unequal and the power factor will have a value somewhere between 0 and 100 percent, depending on the relative values of the resistance, inductance, and capacitance components.

EFFECTIVE VALUES OF VOLTAGE AND CURRENT

As mentioned earlier, when an electric current flows through a circuit, heat is generated or produced in the circuit. The rate at which heat is produced in the circuit is equal to $I^2R;$ where I is the current in amperes and R is the resistance

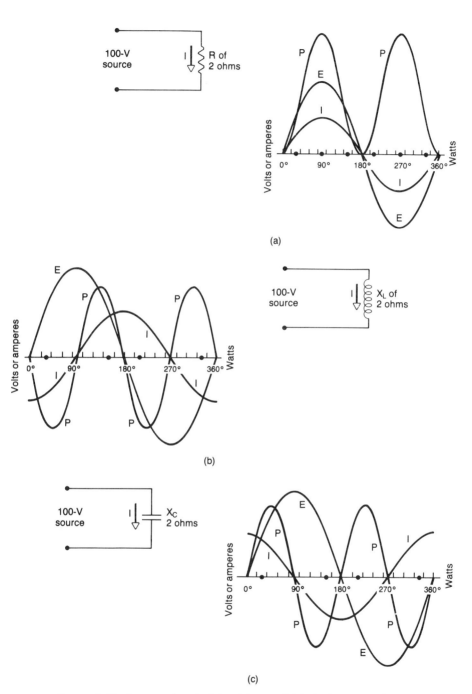

Figure 6.10 Power values in (a) resistance; (b) inductance; and (c) capacitive circuits.

in ohms. This heating effect takes place when either a direct (nonalternating) or an alternating-current flows.

In an alternating-current circuit, the rate at which heat is being generated is constantly changing because the current is constantly changing. In a direct-current circuit, when a current flows, the rate at which heat is developed is constant (for a fixed value of current) because the direct-current remains constant in value. For example, an alternating-current which reaches a maximum value of 100 amperes would not have the same heating effect as a direct-current of 100 amperes. To produce the same heating effect, the alternating-current would have to be of such value that it reaches a maximum value of 141.4 amperes. Such a current varying according to a sine wave between zero amperes and a maximum of 141.4 amperes has the same effective heating value as a direct (constant) current of 100 amperes and, hence, its "effective" value is said to be 100 amperes.

The same effective value is used for voltages. Thus an alternating (sine wave) voltage varying between zero volts and 141.4 volts is said to have an "effective" value of 100 volts.

Both alternating-currents and voltages are usually expressed in terms of effective values. The ratio between the effective and maximum values is illustrated in Figure 6.11. This ratio can be expressed in either of two ways.

$$\frac{\text{Effective value}}{\text{Maximum value}} = \frac{0.707 \times \text{Maximum value}}{1.414 \times \text{Effective value}}$$

Thus, if one value is known, the other can be easily determined.

The maximum value is an instantaneous value. Any other instantaneous values for other instants during the cycle can be determined by scaling off values from the curve of a sine wave.

Alternating-current instruments, such as ammeters and voltmeters, are calibrated to indicate effective values.

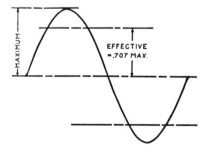

Figure 6.11 Ratio between maximum and effective values of sine wave voltage or current.

BASIC ELECTRIC CIRCUIT

Referring to Chap. 1, the basic circuit for transmitting electrical energy consists of two paths or conductors, one sending and the other a return. This holds for both alternating-current and direct-current systems. In direct-current the voltage applied is continuous and the current flows in one direction continuously. In alternating-current the voltage applied rises from zero to a maximum and drops back to zero, then reverses direction and goes through the same variations; a graph of this pattern is called a "sine wave" (Figure 6.12). Current flows back and forth in a push-pull fashion with each half of the cycle doing productive work. In the United States, systems operate at 60 cycles per second, that is, there are 60 of these push-pull flows each second. In other parts of the world, many systems operate at 50 cycles per second.

Ground Neutral

One such two-wire circuit (Figure 6.13), whether alternating- or direct-current, can carry a certain prescribed amount of power. For safety reasons, usually one conductor — the return — is grounded, that is, connected to earth.

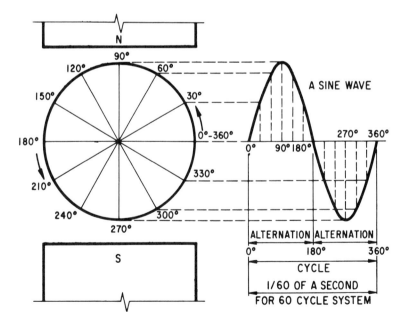

Figure 6.12 Generating the sine wave.

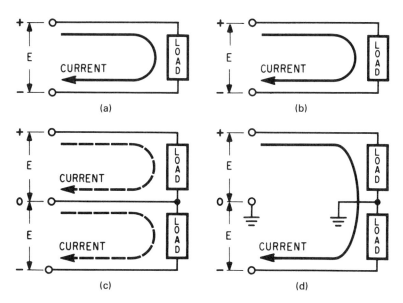

Figure 6.13 Effects of ground/neutral connections. (a) A schematic two-wire dc transmission line where "E" is voltage from the line to the supporting structure. (b) Schematic dc transmission line using earth return. (c) Three-wire dc system where current is cancelled in the center leg. (d) Schematic ± E dc grounded transmission line with twice the capacity of (a) and some insulation and conductor size (current flow).

The voltage applied, then, can be specified as so many volts above or from ground. If additional amounts of power are desired to be transmitted, a second similar circuit can be added. However, in the latter circuit, the voltage applied can be in the reverse direction of the first, that is, so many volts below ground. If the first voltage applied is labeled positive, then the second can be labeled negative. It will be noted that both circuits have one conductor grounded, hence these two ground, or return, conductors can be connected together. It will further be observed that the current in one return circuit opposes that of the other circuit; the common return — or neutral — conductor will now carry only the difference in magnitude of these two currents. The neutral conductor, therefore, need not be as large as either of the main conductors, resulting in economy. Further, if both circuits carried exactly equal currents, the current in the neutral would be zero, and the neutral conductor could be eliminated, resulting in further economy. The voltage difference between both main conductors will now be double that for one. However, since each conductor carries insulation for its original voltage, the insulations existing between both conductors will be sufficient to sustain the double voltage.

POLYPHASE SYSTEMS

For alternating-current systems, the power supplied can be delivered over more than two circuits, as described. These are known as polyphase (or multiphase) circuits. In a single-phase circuit, only one phase or set of voltages of sine-wave form is applied to the circuit and only one phase of sine-wave current flows in the circuit. In polyphase circuits, two or more phases or sets of sine-wave voltages are applied to the different portions of the circuits and a corresponding number of sine-wave currents flow in those portions of the circuits. The different portions of the polyphase circuits are usually called the "phase." Studies indicate an economic and practical arrangement is to limit this number to three, and this is referred to as a three-phase circuit. Practically all alternating-current transmission systems are three-phase and have three-current carrying conductors.

The phases are usually lettered to identify them, as the "A" phase, "B" phase, and "C" phase. The voltages applied to the separate phases of the circuit are correspondingly referred to as the "A-phase voltage," the "B-phase voltage," etc. The phase currents are similarly identified.

In a three-phase circuit, the alternating pulses of electricity are displaced one from another by 120°, that is, one complete cycle is considered to represent 360°. Then, if the voltage in one conductor (A) starts at zero, the voltage in a second conductor (B) with reference to the first will begin its cycle 120° later, and that in the third conductor (C) with reference to the first will begin 240° later (Fig. 6.14). The fundamental principles of the flow of alternating-currents are the same whether applied to single-phase or polyphase circuits.

The voltages for polyphase systems are supplied from polyphase (multiphase) generators, each phase of voltage generated in a separate coil (or coils connected in parallel), the separate coils being arranged for connection in different ways to form the polyphase system. Two commonly used methods of connecting the coils of three-phase generators to supply three-phase transmission systems are shown in Figure 6.15. One method employs the "delta" connection, the other the "wye" or "star" connection.

Delta and Wye Connections

The voltages between the terminals or conductors of the two types of three-phase systems are shown in Figure 6.16. Note the voltages between the terminals are the same for both systems. In the delta system, the full (terminal) voltage is imposed on the coils of the generator or motors whose coils are also connected in delta and which may be connected to the delta system. In the wye system, the voltage imposed on these coils is only 0.866 that of the delta system. The common connection of the three phases in the wye system may be connected to a neutral or fourth conductor, or may be left as is to "float." The

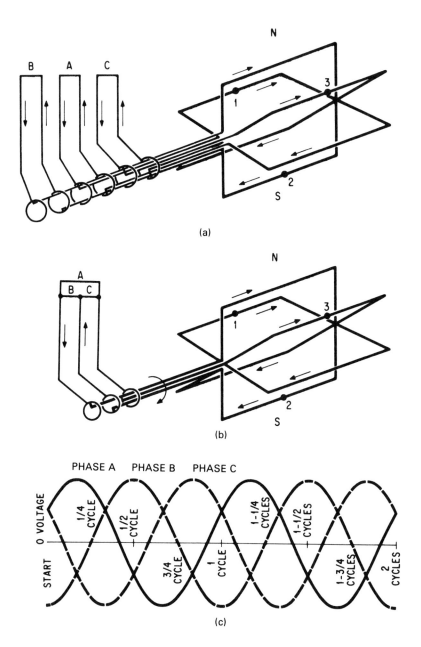

Figure 6.14 A three-phase generator. (a) Three single-phase generators mounted on the same shaft. (b) A three-phase generator. (c) Curves showing voltages in a three-phase generator.

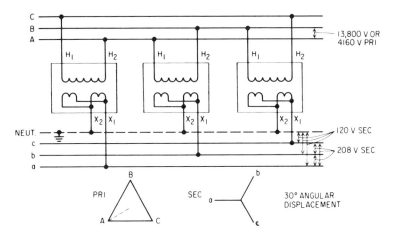

Figure 6.15 Methods of connection of generator (or motor) coils for three-phase systems.

shield wire may also be the fourth wire and carries the unbalance of one or more such three-phase circuits. In some cases, it is also connected to or serves as the neutral conductor for distribution circuits that may be installed on the same supporting structures.

While the coils connected in the delta system are subjected to higher voltages than those in the wye system, as there is no ground connection, one or more grounds (faults) on one phase only will not affect the operation of the system. A second ground on another phase constitutes a short circuit. In the wye system, while the coils have a lower voltage imposed on them, if the system is grounded, a ground (or fault) on any one phase constitutes a short circuit and may cause the circuit to be deenergized with outages to consumers supplied from that circuit. If the system is not grounded, a ground (or fault) will establish a "common" grounded point, not symmetrical to the circuit's three phases, and the voltages on each phase become distorted—the voltage on the grounded phase may be lower than its nominal value, while the voltages on the others may be higher than the nominal values, possibly causing damage from overvoltage to consumers supplied from those two phases.

Considering the supply of power to a motor or engine, direct-current provides a continuous and even flow, making for a smooth effort. With alternating-current, or push-pull, the flow will build to a maximum and then subside. If all the power is supplied in one circuit or one phase, the resulting effort will be very rough. If it can be supplied in three parts or three phases, each one-third in magnitude but applied three times in a given time, the result is less rough. This is similar to hammering home a nail with one big blow, as against using three blows, each one-third in magnitude; the effect is smoother.

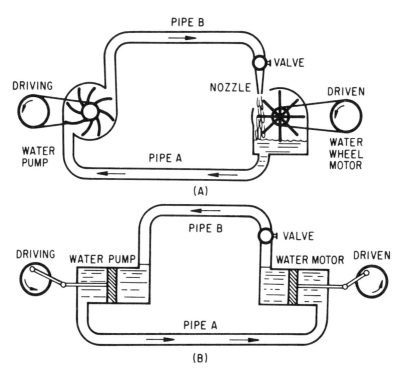

Figure 6.16 The need for using three-phase power. (a) Smooth flow in one direction analogous to direct current power flow. (b) Reciprocating motion analogous to alternating current single-phase power flow. More cylinders would be analogous to three-phase power flow.

Another analogy may be that of a two (large diameter) cylinder engine compared to a six (smaller diameter) cylinder engine in an automobile; the power of both engines is the same, but the latter drives more smoothly than the former. Hence, three-phase power transmission is usually specified in ac systems.

COMPARISON OF ALTERNATING-CURRENT SYSTEMS

A comparison of the various alternating-current systems assuming the same (balanced) load, the same voltage between conductors and the same conductor size is shown in Table 6.1. Using a single-phase, two-wire, circuit as a basis for comparison, the relative amount of conductor, power loss, and voltage drop for the different types are shown.

TABLE 6.1 Comparison of AC Systems

Type of system	Conductor amount	Power Loss	Voltage Drop
Single-phase, 2-wire	1.0	1.0	1.0
Single-phase, 3-wire	1.5	0.25	0.25
Two-phase, 2-wire	1.5	0.50	0.50
Two-phase, 4-wire	2.0	0.50	0.50
Two-phase, 5-wire	2.5	0.50	0.50
Three-phase, 3 wire*	1.5	0.167	0.167
Three-phase, 3 wire**	1.5	0.50	0.50
Three-phase, 4 wire*	2.0	0.167	0.167

*Wye voltage same as single-phase
**Delta voltage same as single-phase
More detail is given in Table 6.2, Table of Transmission Efficiencies.

REVIEW

☐ Electrical pressure is measured in volts, current in amperes, resistance in ohms, power in watts, energy in watt-hours.

☐ The relationship between these quantities

$$\text{Ohm's Law: Current (amperes)} = \frac{\text{Pressure (volts)}}{\text{Resistance (ohms)}}$$

$$\text{Power (watts)} = \text{Pressure (volts)} \times \text{Current (amperes)}$$

$$\text{Energy (watt-hours)} = \text{Power (watts)} \times \text{Time (hours)}$$

☐ Electricity requires a complete circuit to flow. The two kinds of circuits are the series circuit and the multiple or parallel circuit; the series-parallel circuit is a combination of the two.

☐ In a series circuit, the same current flows in each of the components. In a multiple circuit, the voltage across each of the components is the same.

☐ Power is the rate of expending energy. Energy is the expenditure of power over a period of time.

☐ Current in a wire produces heat and a loss or drop in voltage.

☐ Inductance is the obstruction to the flow of current in an ac circuit caused by magnetic lines of force cutting it. These lines of force may be produced by the current flowing in the conductor itself, or by an adjacent conductor. The first is called self-inductance, the latter mutual inductance. They are expressed in ohms.

TABLE 6.2 Transmission Efficiencies[*]

Type of Circuit	Current in Conductor	Power Loss (I^2R) in Conductor	Voltage Drop (IR)
1. Single-phase, 2-wire 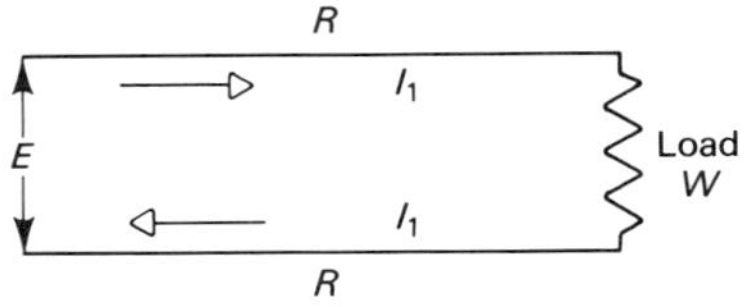	I_1	$I_1^2 \times 2R = 2I_1^2R$	$I_1 \times 2R = 2I_1R$
2. Single-Phase, 3-wire 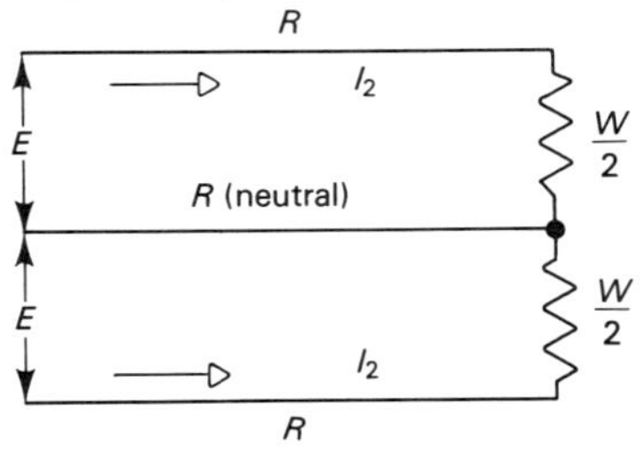	$I_2 = \frac{1}{2}I_1$	$I_2^2 \times 2R = (\frac{1}{2}I_1)^2 \times 2R = \frac{1}{4}(2I_1^2R)$	$I_2 \times R = \frac{1}{2}I_1 \times R = \frac{1}{4}(2I_1R)$
3. Two-Phase, 4-wire 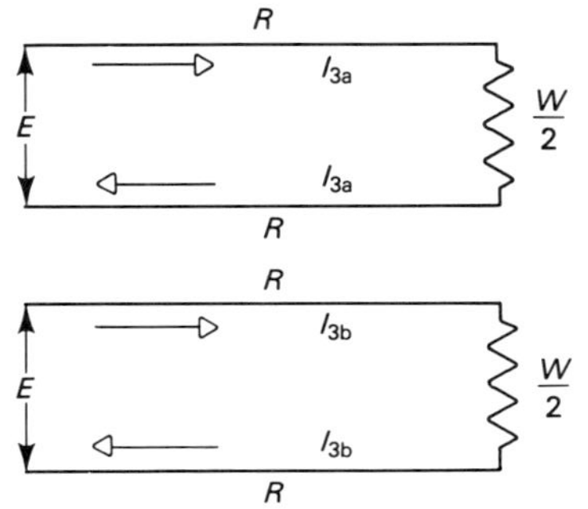	$I_{3a} = \frac{1}{2}I_1$ $I_{3b} = \frac{1}{2}I_1$	$I_3^2 \times 2R \times 2 = 4I_3^2 \times R = 4(\frac{1}{2}I_1)^2 = \frac{1}{2}(2I_1^2R)$	$I_3 \times 2R = 2I_3R = 2(\frac{1}{2}I_1)R = \frac{1}{2}(2I_1R)$

Circuit diagrams (left column):

1. Single-phase, 2-wire: Load W, R, I_1, R, E

2. Single-Phase, 3-wire: $\frac{W}{2}$, $\frac{W}{2}$, R, I_2, R (neutral), I_2, E, E

3. Two-Phase, 4-wire: $\frac{W}{2}$, $\frac{W}{2}$, R, I_{3a}, I_{3a}, R, R, I_{3b}, I_{3b}, R, E, E

$$I_4 = \tfrac{1}{2}I_1$$

$$
\begin{aligned}
I_4^2 &= 2R + \left(\sqrt{2I_4}\right)^2 R \\
&= I_4^2(2R + 2R) \\
&= 4I_4^2 R \\
&= 4(\tfrac{1}{2}I_1)^2 R \\
&= \tfrac{1}{2}(2I_1^2 R)
\end{aligned}
$$

$$
\begin{aligned}
I_4R + \sqrt{2}I_4R &= I_4R + 1.41\,I_4R \\
&= 2.42\,I_4R \\
&= 2.42(\tfrac{1}{2}I_1)R \\
&= 1.21 \times I_1R \\
&= \text{approx } \tfrac{1}{2}(2I_1)R
\end{aligned}
$$

Since loads are balanced, no current will flow in the fifth or neutral wire. Hence, current, power loss, and voltage drop will be the same as for the Two-Phase, 4-wire system; refer to item 3 above.

$$I_6 = \frac{I_1}{3}$$

$$
\begin{aligned}
I_6^2 \times R \times 3 &= \left(\frac{I_1}{3}\right)^2 \times R \times 3 \\
&= \frac{1}{3}I_1^2 R \\
&\text{or } \frac{1}{6}(2I_1^2 R)
\end{aligned}
$$

$$
\begin{aligned}
I_6 \times R = \frac{I_1}{3} &= R \\
&= \frac{1}{6}(2I_1R)
\end{aligned}
$$

4. Two-Phase, 3-wire

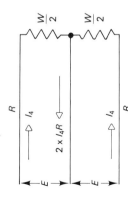

5. Two-Phase, 5-wire

6. Three-Phase, 3-wire — Y

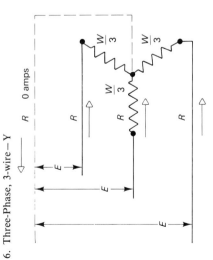

TABLE 6.2 Continued.

Type of Circuit	Current in Conductor	Power Loss (I^2R) in Conductor	Voltage Drop (IR)
7. Three-Phase, 3-wire	$I_7 = \dfrac{I_1}{\sqrt{3}}$	$I_7^2 \times R \times 3$ $= \left(\dfrac{1}{\sqrt{3}}\right)^2 \times R \times 3$ $= I_1^2 R$ or $\frac{1}{2}(2I_1^2 R)$	$I_7 \times R = \dfrac{I_1}{\sqrt{3}} R$ $= \dfrac{1}{2\sqrt{3}}(2I_1 R)$ * *For comparison, this value should be multiplied by $\sqrt{3}$ because line to neutral voltage in this case is only $\dfrac{1}{\sqrt{3}}$ of E assumed in case 1. Therefore, comparative voltage drop $= \frac{1}{2}(2I_1 R)$

8. Three-Phase, 4-wire – Y

Since loads are balanced, no current will flow in the fourth or neutral wire. Hence, current, power loss, and voltage drop will be the same as for Three-Phase, 3-wire – Y system; refer to item 6.

*Based on fixed, balanced loads at power factor of 1.0 and same wire size.

☐ Capacitance is the obstruction to the flow of current in an ac circuit caused by electrostatic fields set up by adjacent conductors; it is expressed in ohms.

☐ Inductance and capacitance in an ac circuit cause the current and voltage to be displaced from each other and not to act together when producing power.

☐ Impedance is the net result of the action of resistance, inductance and capacitance in an ac circuit; it is expressed in ohms.

Ohm's Law for ac becomes:

$$\text{Current (amperes)} = \frac{\text{Pressure (volts)}}{\text{Impedance (ohms)}}$$

$$\text{Power (watts)} = \text{Pressure (volts)} \times \text{Current (amperes)} \times \text{Power factor (percent)}$$

☐ Power factor is the ratio of real power to apparent power; it is expressed in percent.

☐ The effective value of an ac sine-wave voltage or current is 70.7 percent of the maximum value of the sine wave and is equal to the effective value of dc voltage or current values.

☐ The basic circuit for both a dc or an ac circuit consists of two wires, a sending one and a return.

☐ In an ac circuit, power may be delivered over two or three circuits or phases whose voltages are displaced electrically by 120° for three-phase circuits.

STUDY QUESTIONS

1. What are the units of electrical pressure, current, and resistance?
2. Express the relationship between the three quantities in a simple electric circuit. What name is given to this relationship?
3. What are the two basic types of electric circuits? How do they differ?
4. What is the difference between power and energy?
5. What two factors must be considered in transmitting power a great distance?
6. How may a voltage be induced in a conductor? On what factors does the magnitude of the induced voltage depend?
7. What is meant by inductance? What is meant by capacitance? What effect do they have on the relationship between voltage and current in a circuit?

8. What is meant by impedance? How is Ohm's Law affected by alternating-currents? What is meant by resonance?

9. What is meant by power factor in an ac system? How is it expressed?

10. What is meant by a single-phase circuit? By a polyphase circuit? What are the advantages and disadvantages of each?

Environmental Considerations

Many environmental considerations are taken into account in the planning, design, construction, maintenance, and operation of transmission lines and associated substations. A few examples follow to illustrate the scope and magnitude of such considerations.

SAFETY

Overall, perhaps the prime considerations are the measures taken to protect people from injury. In addition to maintaining certain minimum clearances of hazardous objects from the general public, other measures include: barriers, fences and walls around potentially dangerous areas; locks on doors, gates, operating handles and other access facilities; interlocks on critical equipment and devices to prevent incorrect operation and accidental energization and deenergization of facilities that could endanger workers and the public; grounding of fences, towers and other metallic structures accessible to the general public and that could be accidentally energized; smoke and fire alarms to warn the public of impending possible danger; measuring devices installed throughout a region to measure pollution of the surrounding atmosphere; and so on.

ESTHETICS AND APPEARANCE

Location of transmission lines are rerouted to avoid areas of particular interest, such as, national monuments or regional parks, areas of population density or recreational importance, areas where people congregate for special events, playgrounds, buildings and other structures of historical interest, for the preservation of the integrity of national resources and ecological preserves, including animal sanctuaries and forest reserves.

Much attention has been directed toward making transmission lines and associated substations less obtrusive and more pleasing (or less displeasing) to observers. Poles and support structures have been gracefully designed and colors are used to have them blend with surroundings, with similar treatment applied to insulators and equipment mounted on the structures. For lower voltage lines up to 69 kV, lines are constructed without cross-arms to improve appearances (see Chap. 1).

Rights-of-way are cleared of unsightly growth and landscaping is provided in some areas; in other areas, flowers and other plants are used to create pleasant garden effects.

Substations are landscaped to conceal or beautify the enclosures of equipment. In some instances, the station is contained within structures made to conform and blend in with the neighborhood.

When other means have proven impractical, lines are placed underground for short distances through the areas affected, including cable adits and exits from substations.

POLLUTION

Sight is not the only sense to which attention is paid. Steps are taken to eliminate or abate disagreeable sounds. Acoustic barriers are installed, generally at substations, to keep annoying sounds confined within limited areas. In some instances, devices producing sound frequencies to cancel those produced by energized equipment are installed to lessen this form of nuisance.

Possible pollution from oil spills of ground and underground water sources from equipment (transformers, circuit breakers, etc.) is prevented by the design and construction of dams and barriers around such equipment).

ECOLOGY

The preservation of wild life refuges and sanctuaries, including endangered species, by the rerouting of facilities has been discussed above. Other provisions are made to protect some wild life from self-destruction (as well as to prevent interruption of service). These include shields placed on the lower parts of poles, towers, and other structures to keep such animals from contact with energized conductors, and similar guards installed on equipment bushings to serve the same purpose. In some instances, usually where ospreys, eagles, and other birds with wide wing spreads exist, nests are mounted on separate structures a short distance away from the transmission lines. These are meant to keep such birds from nesting on or in the line structures where they may come in contact with energized conductors.

HAZARDOUS MATERIALS

In another field, hazardous materials are avoided initially, or where they may exist, steps taken to eliminate them. Where insulating oils and askarels contain polychlorinated biphenyl (PCB), steps are taken to replace the fluids (where necessary the entire equipment itself), or clarify them. PCB is a substance alleged to cause cancer to its handlers. Similarly, where asbestos insulation or noise barriers contain asbestos, steps are taken to remove this material or replace it with other nonhazardous materials (Chap. 4).

ELECTROMAGNETIC RADIATION

In the case of higher voltage transmission lines, research continues on the effect of their strong magnetic fields on human, animal, and plant life in their vicinity. As is known, a voltage is induced in a conductor cut by a magnetic field, its magnitude in part depending on the strength of the magnetic field. People, animals, and plants constitute conductors of sorts and have voltages induced in them that cause currents to flow within them as "eddy" currents, and between them and the ground or other objects with which they may come in contact.

   The effect these currents have on the biological system, and especially on the nervous system, appears to be minimal or nonexistent, although the length of time of such exposure may play an important part. Meanwhile, tentative standards are being drawn calling for minimum widths of rights-of-way in accordance with the voltage of the transmission lines. For example, some states have legislated a minimum width of right-of-way of 350 feet (or 171 metres) with additional width provided to maintain a maximum magnetic field strength of 1.6 kV per metre from conductor to the area or structure in its vicinity applied to the shortest distance between them.

   These observations apply to both alternating-current and direct-current lines. Magnetic fields produced by the former alternate in direction according to the frequency of the circuit; this constant to-and-fro motion produces alternating-current on whatever bodies that are cut by the magnetic fields. In direct-current transmission, the magnetic fields produced are stationary (except when first energized and again when deenergized); however, objects in motion in their vicinity cut the magnetic field, producing a direct-current within them. Tentative "wire codes" have been established by some utilities that correlate the number of wires and their diameters and voltages to specific "safe" distances from vulnerable objects. These codes are modified by other factors, including pollution effects on the lines, density of people that may be in the vicinity of the line, traffic in the area, the shielding effect of other structures, and other factors that may influence possible current flow.

* * *

These are only a few of the measures taken to improve and maintain the quality of the environment that may be influenced by transmission lines. They are an indication, however, of the scale and scope of the considerations taken into account that affect transmission lines from the planning stages to their final construction and operation. The accompanying figures are some examples of the subject matters under discussion.

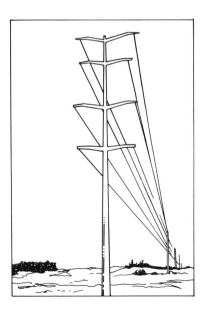

Figure A.1 345kV single shaft double circuit.

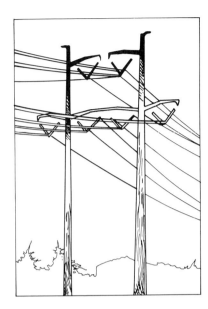

Figure A.2 500kV H-frame double circuit.

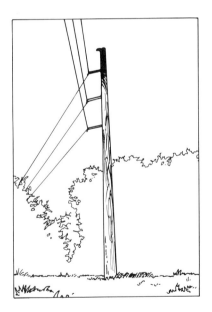

Figure A.3 Light duty steel pole.

Figure A.1, A.2, and A.3: Three modern designs of transmission structures. Members are made of laminated wood, steel, or aluminum, and are colored to fit in with neighboring structures. They are designed to be inconspicuous as well as practical. (*Courtesy of FL Industries, Minn.*)

Figure A.4 Atlantic Beach substation. Low profile substation has fencing and shrubs typical of the neighborhood. Structures are painted in colors defining their type and function, with special emphasis on hazardous areas; major equipment is colored to blend in with the environment—to be as inconspicuous as possible. (*Courtesy of Long Island Lighting Company.*)

Figure A.5 Seaport substation in Manhattan. Walls around equipment made to look like other industrial buildings in the area. (*Courtesy of Consolidated Edison Company.*)

Figure A.6 Green Acres substation. Heavy landscaping matches that typical of the area. Growth is monitored to maintain appearance, shielding structures from view as much as possible. (*Courtesy of Long Island Lighting Company.*)

Figure A.7   138kV lines. (left) 1987; (right) circa 1937. (*Courtesy of Long Island Lighting Company.*)

Appendix B # Guide for Uprating Transmission Structures For Higher Operating Voltages

Transmission lines can have their capacities uprated to accommodate increased loading to utilize existing rights-of-way. This may be accomplished by installing larger conductors on existing structures, increasing operating voltages, or both. In general, uprating existing lines creates fewer environmental, economic, and public relations problems than obtaining rights-of-way for new facilities. In such uprating, however, reconstruction and replacement of structures as well as revisions and beautification of existing rights-of-way may be necessary.

The guide details the handling of all of the elements in the transmission line to be uprated. It lends itself admirably to working out the details for obtaining and preparing rights-of-way as well as those associated with the construction of a new transmission line. Hence, this guide was selected to accomplish both purposes, like the proverbial killing of two birds with one stone.

Standards and methods employed are those of the Rural Electrification Administration (REA) of the U.S. Department of Agriculture, to whom thanks are due for extending this courtesy. While the standards and methods of REA may differ in some details from those of other public and private utilities, they do represent a consensus of the utility industry and should prove useful for the purposes indicated. This is especially appropriate when it is remembered that an important part, if not a major part, of REAs activities include the planning, design, construction, maintenance and operation of high voltage, large capacity, very long cross country transmission lines.

Additional information, specifications, and other material may be requested and obtained from REA offices.

CASE STUDY: 69 kV TO 115 kV LINE UPRATING

Introduction

This analysis provides a brief summary of a 69 kV to 115 kV transmission-line uprating study. The original 69 kV line was constructed on the standard REA TS-1 single-pole structure (Fig. B.1). The uprated 115 kV line structure was constructed on the TH-1AM structure (Fig. B.2).

Analysis

The original 69 kV line was designed and constructed in 1951–1952 under the safety guidelines of the 4th Edition of the National Electrical Safety Code (NESC) and the then existing REA Transmission Line Design Guides. The uprated 115 kV line was designed in 1979 in accordance with the 1977 Edition of the NESC and the 1972 REA Bulletin 62-1 and July 1978 "File With" REA Bulletin 62-1 vertical clearance requirements.

A tabulation of the original and uprated design criteria is provided below.

Original Design Summary	Uprated Design Summary
Conductor: #1/0 ACSR 6/1	477 MCM 26/7 ACSR
Shield wire: 3/8″ H.S. Steel	3/8″ H.S. Steel
Loading: 1/2″ Ice & 4# Wind @ 0° F	1/2″ Ice & 4# Wind @ 0° F
Ruling span: 500′	380′
Voltage: 69 kV	115 kV
Basic poles: 50′Cl.3	(50′Cl.3 & 55′Cl.3) 1 ea.

The original 69 kV line had a design ruling span of 500 feet as compared with an actual ruling span of 380 feet based on structure locations. A check of several contemporary line projects revealed a similar practice.

Summation of moments at the groundline because of wind and ice loads revealed that the larger 477 MCM 26/7 conductor and two 3/8" H.S. Steel shield wires could be installed on the uprated TH-1AM tangent structure and provide a 4.0 safety factor under NESC Heavy Loading criteria. The maximum allowable sum of adjacent spans for several typical structure heights are on the following pages.

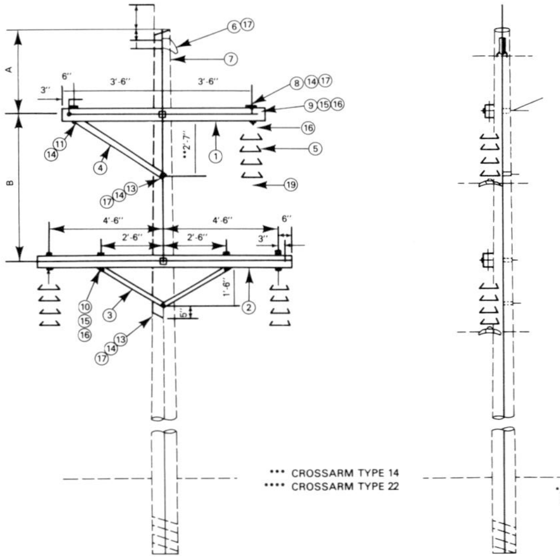

*** CROSSARM TYPE 14
**** CROSSARM TYPE 22

*As required. See Drg. TM-1

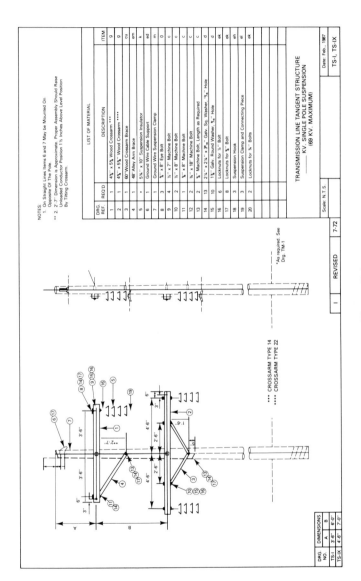

NOTES:
1. On Straight Lines Items 6 and 7 May be Mounted On Opposite Of The Pole.

** 2. 2'-7" Dimension is Approximate. Proper Assembly Should Raise Unloaded Conductor Position 1½ Inches Above Level Position By Tilting Crossarm.

LIST OF MATERIAL

DRG. REF.	REQ'D	DESCRIPTION	ITEM
1	1	4⅝" x 5⅝ Wood Crossarm ***	g
2	1	4⅝" x 5⅝" Wood Crossarm ****	g
3	1	60" Wood Crossarm Brace	cu
4	1	48" Alley Arm Brace	em
5	•	5¾" x 10" Suspension Insulator	k
6	1	Ground Wire Cable Support	ed
7	1	Ground Wire Suspension Clamp	m
8	3	⅝" x 8" Eye Bolt	0
9	4	½" x 7" Machine Bolt	c
10	2	½" x 8" Machine Bolt	c
11	1	⅝" x 8" Machine Bolt	c
12	2	¾" x 18" Machine Bolt	c
13	2	⅝" Machine Bolt, Length as Required	c
14	13	2¼" x 2¼" x 3/16" Galv. Sq. Washer, 13/16" Hole	d
15	10	1⅜" Galv. Round Washer, 9/16" Hole	d
16	6	Locknuts for ½" Bolt	ek
17	8	Locknuts for ⅝" Bolt	ek
18	3	Suspension Hook	eh
19	3	Suspension Clamp and Connecting Piece	el
20	2	Locknuts for ¾" Bolts	ek

TRANSMISSION LINE TANGENT STRUCTURE
_____ KV. SINGLE POLE SUSPENSION
(69 KV. MAXIMUM)

DRG. NO.	DIMENSIONS A	DIMENSIONS B
TS-I	3'-6"	6'-0"
TS-IX	4'-6"	7'-0"

I	REVISED	7-72	Scale: N.T.S.		Date: Feb., 1987
					TS-I, TS-IX

Figure B.1

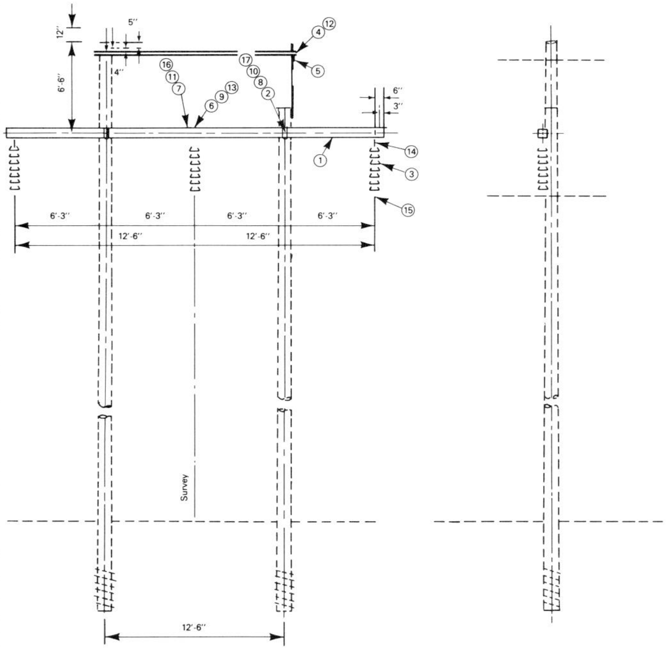

*As required. See Drg. TM-1

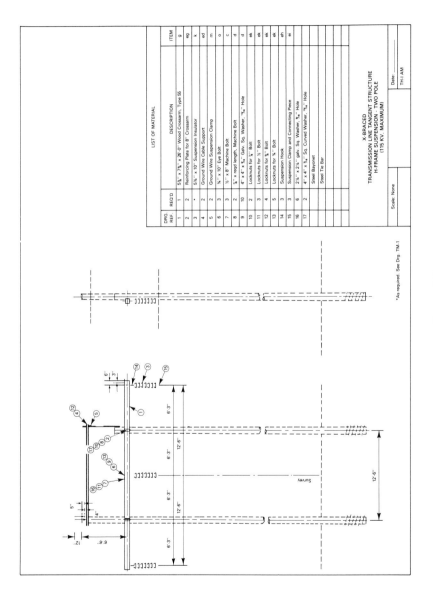

LIST OF MATERIAL

DRG. REF.	REQ'D	DESCRIPTION	ITEM
1	1	5⅜″ x 7⅝″ x 26′-0″ Wood Crossarm, Type 55	g
2	2	Reinforcing Plate for 8″ Crossarm	eg
3	*	5¾″ x 10″ Suspension Insulator	k
4	2	Ground Wire Cable Support	ed
5	2	Ground Wire Suspension Clamp	m
6	3	¾″ x 10″ Eye Bolt	o
7	3	½″ x 8″ Machine Bolt	c
8	2	⅞″ x reqd length, Machine Bolt	d
9	10	4″ x 4″ x $\frac{3}{16}$″ Galv. Sq. Washer, $\frac{13}{16}$″ Hole	d
10	2	Locknuts for ⅞″ Bolt	ek
11	3	Locknuts for ½″ Bolt	ek
12	4	Locknuts for ⅝″ Bolt	ek
13	5	Locknuts for ¾″ Bolt	ek
14	3	Suspension Hook	eh
15	3	Suspension Clamp and Connecting Piece	ei
16	6	2¼″ x 2¼″ galv. Sq. Washer, $\frac{9}{16}$″ Hole	
17	2	4″ x 4″ x $\frac{3}{16}$″ Sq. Curved Washer, $\frac{15}{16}$″ Hole	
		Steel Bayonet	
		Steel Tie Bar	

X-BRACED
TRANSMISSION LINE TANGENT STRUCTURE
H-FRAME SUSPENSION - TWO POLE
(115 KV. MAXIMUM)

Scale: None		Date: ________
		TH-I AM

Figure B.2

TH-1AM (Ht. and Class)		Max. Sum of Adjacent spans (SF = 4.0)
50	3	1151 feet
55	3	1136 feet
60	3	1129 feet

The single crossarm (Type 55) of the TH-1AM structure has a maximum vertical span limitation of 822 feet under NESC H.L. conditions and a 4.0 safety factor.

The allowable sum of adjacent spans and the vertical span allowed by the cross-arm strength, permitted existing poles and pole locations to be used in

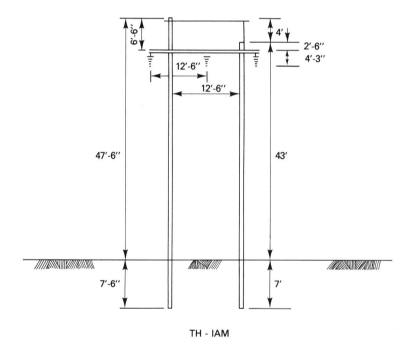

TH - IAM

Figure B.3 NORMAL TANGENT SPAN ON LEVEL GROUND

STRUCTURE TYPE: MODIFIED H-FRAME
POLE: 50' CLASS 3
CONDUCTOR: 477 26/7 ACSR HAWK

HEIGHT OF LOWEST COND. ABOVE GROUND   =   36.25
GROUND CLEARANCE + 1 FT.   =   25.00
SAG IN LEVEL GROUND SPAN   =   11.25

SAG IN 380 FT. RULING SPAN @ MAX. TEMP. FINAL = 8.15

SPAN = SQUARE ROOT OF (11.25 × 380 × 380 ÷ 8.15) = 446.46 FT.

spotting all tangent structures in the uprated line. Additional span lengths could have been attained by installing X-braces or using an assembled cross-arm. The span lengths in the existing line, however, allowed the use of the unbraced, single-piece cross-arm structure described above.

As a general rule, standard REA angle and deadend structures were called for at line angles and deadend points in the uprated line. No attempt was made to redesign or uprate existing 69 kV structure configurations. This rule was employed to assure ample strength and electrical clearances at these critical line locations.

The galloping ellipse patterns were calculated and plotted for the longest individual span in the line (590'). The single loop method of analysis was used and less than 10 percent overlap was detected; therefore, the structure spacing, wire sags, and maximum span length were concluded to be satisfactory.

Inspection/Survey

A vital part of the design procedure in a line uprating is a thorough inspection of the existing line. A competent inspection crew should be assigned to foot patrol and inspect the entire line length. The inspection crew should use copies of the existing line plan and profile drawings. They should perform many checks including the following:

Verify the elevation of the elevation profile.

Verify the height and groundline circumference of each existing structure.

Verify the height and survey station of each utility crossing.

Report the location and extent of any right-of-way encroachment (house, barn, mobile home, radio/TV antenna, etc.).

Verify survey station of all highway/road/railroad crossings.

Report any land use changes (original pastureland now under cultivation, pivot irrigation now installed, etc.).

Report any terrain changes (excavations, landfills, stock tanks, terraces, etc.).

Report structure requiring line maintenance (ground or internal rot, damaged pole, etc.).

Line conversion or voltage uprating may warrant a complete new line survey. If the existing transit and level data is relatively recent (less than ten years) and the line route is not in a developing urban area, the existing survey information may be adequate if verified thoroughly by the foot patrol inspection noted above.

An important point to recognize in line uprating is the fact that the

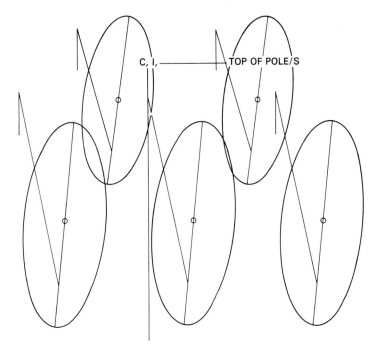

C, I, ──────── TOP OF POLE/S

FOR SPAN LENGTH OF 590 FEET
SINGLE LOOP   LISSAJOUS   ELLIPSE PATTERN
STRUCTURE: MODIFIED H-FRAME

Figure B.4   DATA TABLE FOR WIRE CODEWORD ⅜" H.S. STEEL

ANGLE = 15.672537 AND ANGLE + ANGLE/2 = 23.508806 DEGREES
A + E = 12.197964 FEET
MAGNITUDE OF MAJOR AXIS = 16.247455 FEET
MAGNITUDE OF MINOR AXIS = 6.498982 FEET
B = 3.049491 FEET

POINT OF ATTACHMENT = −3.25 DOWN AND 6.75   FROM CENTER LINE
POINT OF ATTACHMENT = −3.25 DOWN AND −6.75 FROM CENTER LINE

DATA TABLE FOR WIRE CODEWORD HAWK

ANGLE = 11.652755 AND ANGLE + ANGLE/2 = 17.479133 DEGREES
A + E = 18.851455 FEET
MAGNITUDE OF MAJOR AXIS = 19.652519 FEET
MAGNITUDE OF MINOR AXIS = 7.861008 FEET
B = 3.730504 FEET

POINT OF ATTACHMENT = 2.75 DOWN AND 12.5 FROM CENTER LINE
POINT OF ATTACHMENT = 2.75 DOWN AND 0   FROM CENTER LINE
POINT OF ATTACHMENT = 2.75 DOWN AND −12.5 FROM CENTER LINE

centerline of the uprated structure may not be the same centerline of the original structure. The original centerline of the TS-1 structure coincided with the original survey centerline. However, the centerline of the uprated TH-1AM structure was offset 1.91m (6'-3") from the original survey centerline. Because of the flatness of the terrain passed through in this study the offset did not create a real problem. However, a similar offset through heavily wooded or side sloping terrain might have been prohibitive.

Right-of-Way

The original TS-1 single pole 69 kV transmission line required only 15.24m (50 feet) of right-of-way (7.62m [25 feet] either side of the centerline).

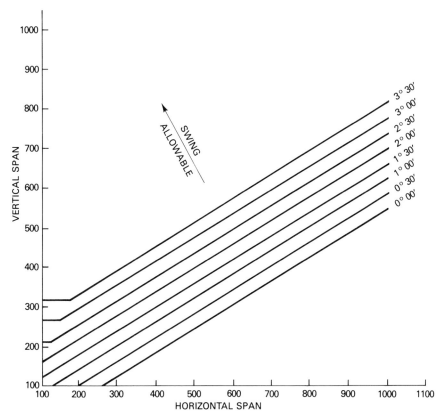

Figure B.5

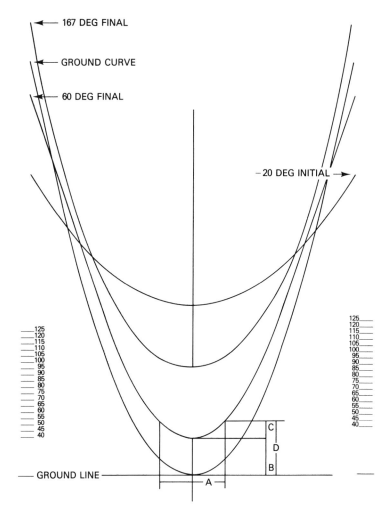

Figure B.6

MODIFIED-H FRAME

477 26/7 ACSR HAWK 380 FT FULING SPAN 115.00 KV HEAVY LOADING
ZONE 50/3 POLE MODIFIED H-FRAME STRUCTURE MAXIMUM DESIGN TEN-
SION = 5819 LBS (29.95↑) 0 DEG INITIAL TENSION = 4080 LBS (21.00↑) UL-
TIMATE STRENGTH = 19430 LB.S

A. LEVEL GROUND SPAN =   446.5
B. GROUND CLEARANCE =   25
C. SAG IN L.G. SPAN =   11.25
D. HT ABOVE GR. OF LOWEST COND =   36.25
E. DIST. FROM POLE TOP TO COND =   6.75

SAG TEMPLATE   SCALE: 40/400

The uprated TH-1AM two pole 115 kV line requires a minimum of 22.86m (75 feet) of permanent right-of-way (9.14m [30 feet] from one side to the original centerline plus 12.19m [40 feet] on the other side to allow for centerline and conductor offset, blow-out and electrical clearance).

Notification

A "notice of intent to construct" is usually required by local regulatory agencies. This type of notification should be given even if it is not required by statute. Information concerning line route, structure configuration, wire size, minimum ground clearances, operating voltage, etc., should be sent to all local utilities, highway departments, county officials, pipeline companies, rural water cooperatives, Federal Aviation Agency, or any other group which may have facilities in the area or be affected by the new construction.

Design Data

The following pages provide a design data summary of the TH-1AM uprated line and structure. One of the main advantages of using the TH-1AM configuration is that it allows an increased ground clearance of 1.68m (5'-6") over the ground clearance allowed by the original TS-1 structure. This provided enough additional ground clearance to allow use of larger conductor as well as voltage upgrading. (An additional 0.46m [1'-6"] of ground clearance could be provided by raising the crossarm to within 0.3048m [1'-0"] of the pole top of the original structure.)

The lightning shield angle of the TH-1AM structure is approximately 28 degrees to the outside phases.

CASE STUDY: TRANSMISSION LINE DESIGN DATA SUMMARY

1. Date: _____April 1979_____
2. Project Designation: _____
3. Line Identification: _____115 kV Uprating_____
    (a) Length, miles: Transmission _____34.5 (approx)_____
    (b) Length, miles: Underbuild: _____ – – – _____
4. *Voltage:*
    (a) Transmission: _____Upgrade from 69 kV to 115 kV_____
    (b) Underbuild: _____ – – – _____
5. Number of Phases on Underbuild: _____

6. *Mechanical Data:*

Item	Transmission Conductor	Overhead Ground Wire	Under-build	Common Neutral
Wire code word	Hawk			
Size (AWG)	447 MCM	3/8″		
Stranding	26/7	7		
Material	ACSR	H.S. Steel		
Diameter (in.)	0.858	0.360		
Area (sq. in.)	0.4356	0.07917		
Weight (lb/ft.)	0.6566	0.273		
Ultimate Strength	19,430#	10,800#		

7. National Electric Safety Code Loading Zone: ____Heavy____

    (a) Ice, (in.) ____1/2____, **(b)** Wind (lb) ____4____

    (c) Fahrenheit ____0°____

8. *Design Data:*

1/2″ ice 4 + .30 wind	Conductors			
	Trans.	O.H.G.W.	Under-build	Common Neutral
Vertical lb/ft	1.501	0.807		
Transverse lb/ft	0.6193	0.4535		
Resultant + K lb/ft	1.934	1.215		
Limiting Tension Conditions (%)				
Initial 0° F/60° F	21.0/13.7	25.0/22.8		
Final 0° F/60° F	18.5/11.4	19.3/16.6		
Maximum	29.95	34.4		

9. Length of span (ft):

    (a) Average (Est.) ____380____, **(b)** Maximum (Est.) ____590____,

    (c) Ruling ____380′____

10. Sag and Tension Chart Nos. ____1-782____, ____1-1330____

    (a) Name of Company: ____Alcoa____

11. Ruling Span, Sags and Tensions:
   (a) *Sags (ft):*

Condition	°F	Transmission		Overhead Ground Wire		Underbuild	
		Initial	Final	Initial	Final	Initial	Final
$1/2$ in. ice $4 + K$ wind	0	5.98	6.01	5.97	5.97		
$1/2$ in. ice, no wind	32	5.83	6.19	4.86	5.06		
Bare, no wind	−20	2.55	2.77	1.69	1.82		
Bare, no wind	0	2.91	3.30	1.83	2.00		
Bare, no wind	32	3.64	4.36	2.08	2.36		
Bare, no wind	60	4.46	5.38	2.37	2.76		
Bare, no wind	90	5.44	6.46	2.76	3.28		
Bare, no wind	120	6.43	7.40	3.23	3.90		
Bare, no wind	167	7.91	8.15	4.15	4.96		
Bare, no wind	212	8.80	8.85	5.13	5.99		

11. Ruling Span, Sags and Tensions: (continued)
   (b) *Tensions (lb):*

Condition	°F	Transmission		Overhead Ground Wire		Underbuild	
		Initial	Final	Initial	Final	Initial	Final
$1/2$ in. ice $4 + K$ wind	0	5819	5781	3712	3712		
$1/2$ in. ice, no wind	32	4652	4387	2999	2881		
Bare, no wind	−20	4642	4288	2911	2712		
Bare, no wind	0	4080	3589	2700	2466		
Bare, no wind	32	3255	2719	2365	2090		
Bare, no wind	60	2660	2205	2079	1788		
Bare, no wind	90	2183	1837	1789	1502		
Bare, no wind	120	1846	1606	1526	1266		
Bare, no wind	167	1501	1459	1189	994		
Bare, no wind	212	1351	1343	961	824		

(c) *Slack Span:*

    (1) Length _____ ft,   **(2)** Design tension _____ lb.

    (3) Initial 60°F sag _____ ft.

    (4) Initial 60°F tension _____ lb.

12. Minimum Conductor Ground Clearance at _____167_____ °F Final sag:

	Nature of Clearance	Transmission (ft)	Underbuild (ft)
(a)	Track rails of railroads	32	
(b)	Public streets and highways	24	
(c)	Areas accessible to pedestrians	19	
(d)	Cultivated field	24	
(e)	Along roads in rural districts	22	
(f)	Additional allowance – sag template	1	
(g)	Communications lines	8	
(h)	Supply lines	6	
(i)	Template cut for a clearance of	25	

13. Base pole:

    (a) Height _____55′ (new) 50′ (existing pole) CL.3_____

    (b) Class _____3_____

    (c) Depth of setting _____7.5′ (new pole) 7.0′ (existing setting depth)

    (d) Type of tangent structure _____TH-1AM_____

14. Crossarm dimensions:

    (a) Width ____$5^3/_8$____ in.,   **(b)** Height ____$7^5/_8$____ in.

15. Structure data:

	Item	Crossarm	Pole
(a)	Shear parallel to grain, lb per sq. in.	1140	1140
(b)	(1) Bending, lb per sq. in.	8000	8000
	(2) Compression, end grain	7420	7420
	(3) Compression, across grain*	910	910
(c)	Species of wood	D. Fir.	SYP
(d)	Preservative	Penta	Penta
	(1) Retention (lb)	0.40	0.50
	(2) Method	Pressure	Pressure

*Stress at proportional limit.

16. Span length limitations:

	Item	Feet
(a)	Normal tangent span on level ground	446
(b)	Maximum sum of adjacent spans with side guys	– – –
(c)	Maximum sum of adjacent spans limited by pole strength (50'–3)	1151
(d)	Maximum vertical span limited by strength of crossarm	822
(e)	Maximum span limited by conductor separation (1) Tangent structure	896
	(2) Tangent to vertical structure	– – –
(f)	Maximum sum limited by underbuild (1) Due to _____	– – –

17. (a) Average number of line angles per mile _____

    (b) Maximum working load for Log Anchor:

        (1) Five foot log ____8,000____ lb.

        (2) Eight foot log ____16,000____ lb.

        (3) Screw Anchors as per manufacurer's recommendation w/SF of 2.

    (c) Guy slope (L/H) ____1:1 Unless Noted____

    (d) Maximum design tension in guy wire $^{3/8''}$ H.S. Steel = 5,400 lb.

    (e) Elevation above sea level:

        (1) Maximum ____4,200____ feet

        (2) Minimum ____3,600____ feet

18.

	Conductor Clearance:	Inches
(a)	Normal support	42
(b)	Minimum to support	26
(c)	Minimum to guy	33
(d)	Climbing or working space for underbuild	– –

19. *Tangent structure conductor configuration: (Check one)*
    Single Pole: Delta ( ), Davit Arms ( ), Steel Arms ( ),
                Horizontal ( ),
    H-Frame: Horizontal (X), Delta ( )

20. *Conductor separation at support*

Distance Between: Locations		(Vertical) Feet	(Horizontal) Feet
(a)	OHGW to upper conductor [1] [2]		
	Upper conductor to middle conductor [1] [2]		
	Upper conductor to lower conductor [1] [2]		
	Middle conductor to lower conductor [1] [2]		
(b)	Upper conductor to upper conductor [2]		
	Middle conductor to middle conductor [2]		
	Lower conductor to lower conductor [2]		
(c)	Horizontal circuit only:		
	Upper center phase to upper right phase	— — —	12.5
	Upper center phase to upper left phase	— — —	12.5
	OHGW or neutral to closest upper conductor	9.3	5.75
	Upper center phase to lower center phase		
	Upper right phase to lower right phase		
	Upper left phase to lower left phase		
(d)	Lowest conductor to closest underbuild phase		

[1] If double circuit single pole, dimensions will be for one side of structure only.
[2] Not applicable for structure with 3-phase circuit in a horizontal plane.

21.

Conductor Separation at Midspan		Separation In Feet	
		Average span	Maximum span
Avg. span = 380 ft, Max. span = 590 ft.			
(a)	Normal—Transmission of OHGW @ 60°F	12.62	16.32
(b)	Normal—Transmission to underbuild @ 60°F	— — —	— — —
(c)	Iced transmission to bare underbuild @ 32°F	— — —	— — —
(d)	Iced OHGW to bare transmission @ 32°F	9.30	8.31
(e)	Ratio of OHGW to conductor @ 60°F Final*	0.513	0.513

*Unitless quantity

22. Wind pressure for insulator side swing:
    (a) Pounds per sq. foot on bare conductor _____ 9 = 0.6435 _____ Lb/Ft
    (b) Pounds per sq. foot on _____ inch iced conductor _____

23. Allowable angle of swing for insulator strings:

Type of structure	No. of ins. units	Maximum	Normal	Minimum	Negative
TH-1AM	7	57.1°	28.5°	– – –	– – –

24. Insulator design loading:
    (a) Suspension insulators
        Class _____ 52–3 _____ (Tangent)
        Maximum load allowed _____ 6,000 _____ lb.
        ANSI M&E rating _____ 15,000 _____ lb.

        Class _____ 52–3 _____ (Angle)
        Maximum load allowed _____ 6,000 _____ lb.
        ANSI M&E rating _____ 15,000 _____ lb.

        Class _____ 52–5 _____ (Deadend)
        Maximum load allowed _____ 10,000 _____ lb.
        ANSI M&E rating _____ 25,000 _____ lb.
    (b) Pin or post insulator maximum cantilever
        design load _____ – – – _____ lb.

25. Phase arrangement: (See Structure drawings and plan and profile notes)

26. Conductor vibration data:
    (a) Should vibration dampers be used to prevent conductor damage?
        NO
    (b) If yes to (a), how many are required? _____
    (c) Type of armor rods _____
    (d) Records of galloping conductor (attach).

27. Weather data:
    (a) Temperature range (°F): Min. _____ −13° _____ Max. _____ 104° _____
    (b) Icing condition: Maximum radial thickness _____ $^1/_2$–$^3/_4$ _____ inches
    (c) Sleet storms: Number of days per year _____
    (d) Maximum hoarfrost on conductor _____ inches
    (e) Maximum height of snow on ground under conductor _____ feet
    (f) Rainfall per year _____ 12 _____ inches

(g) Annual storm days ___45–50/year___

(h) Atmospheric contamination (describe below in remarks):

(i) Wind velocity: Annual mean speed = 8 mph

(1)

Strongest (5 minute) maximum for year:	MPH
a. 19 _____	
b. 19 _____	
c. 19 _____	
d. 19 _____	

(2)

Percent time wind blows:	Percent
a. 1 to 3 MPH	
b. 3 to 8 MPH	
c. 8 to 15 MPH	
d. 15 to 30 MPH	

(j) Prevailing wind direction ___West to East (Annual mean)___

28. Average Terrain of Right-of-Way:

(a) Describe briefly:

(a) Flat to gently rolling terrain. Mostly pasture with some flood irrigation farmland.

(b) Character of soil: ___Sandy___

(c) Soil bearing value _____ lb/sp. in. average

(d) Number of rock holes ___NONE___

(e) Construction in swamps (if any) ___NONE___ miles

(f) Depth of ground water level _____ feet

(g) Frost line depth _____ feet

(h) Remarks:

(i) Accessibility by roads, etc. (describe):
Centerline is accessible from established county roads.

    (j) Width of right-of-way ————————————————— feet
    (k) Estimated structure footing resistance ————10———— ohms
    (l) Right-of-way clearing ————(approx.) one———— miles
  (m) Underground corrosion (describe):
        None reported.

    (n) Underground stray currents (describe):
           None reported.

Items Normally Required for Transmission Line Specifications:

	Item	Check
(a)	Description of units, specifications and drawings for transmission line construction	
(b)	Transmission line design data summary	
(c)	Vicinity and key maps	
(d)	Plan and profile drawings	
(e)	Arrangement of guys	
(f)	Arrangement of lines and overhead ground wire around substations	
(g)	Phasing diagram	
(h)	Single line diagram	
(i)	Sag template	
(j)	Tables for stringing sags and tensions	
(k)	Structure list	
(l)	Insulator swing charts	
(m)	Comparative cost data per mile	
(n)	Justification for underbuild	
(o)	Copy of detailed design calculations	

Sag and Tension Data

Prepared for Heavy Loading Zone

3/8" H.S. Steel

Area = 0.07917
Ultimate = 10,800 Lb

Span = 380.0

	Design Points		Final			Initial		
Temp.	Ice	Wind	Sag	Ten.	P/Ult.	Sag	Ten.	P/Ult.
0	½ in.	4 + .30	5.97	3712	34.37	5.97	3712	34.37
32	½ in.	No wind	5.06	2881	26.68	4.86	2999	27.77
−20	Bare	No wind	1.82	2712	25.11	1.69	2911	26.95
0	Bare	No wind	2.00	2466	22.83	1.83	2700	25.00
60	Bare	No wind	2.76	1788	16.55	2.37	2079	19.25
32	Bare	No wind	2.36	2090	19.35	2.08	2365	21.90
90	Bare	No wind	3.28	1502	13.91	2.76	1789	16.56
120	Bare	No wind	3.90	1266	11.72	3.23	1526	14.13
140	Bare	No wind	4.34	1136	10.52	3.60	1370	12.69
167	Bare	No wind	4.96	994	9.21	4.15	1189	11.01
212	Bare	No wind	5.99	824	7.63	5.13	961	8.90

Sag and Tension Data

Prepared for Heavy Loading Zone

477 26/7 ACSR Hawk

Area = 0.43560
Ultimate = 19,430 Lb

Span = 380.0

	Design Points Creep is a Factor		Final			Initial		
Temp.	Ice	Wind	Sag	Ten.	P/Ult.	Sag	Ten.	P/Ult.
0	½ in.	4 + .30	6.01	5781	29.75	5.98	5819	29.95
32	½ in.	No wind	6.19	4387	22.58	5.83	4652	23.94
−20	Bare	No wind	2.77	4288	22.07	2.55	4642	23.89
0	Bare	No wind	3.30	3589	18.47	2.91	4080	21.00
60	Bare	No wind	5.38	2205	11.35	4.46	2660	13.69
32	Bare	No wind	4.36	2719	13.99	3.64	3255	16.75
90	Bare	No wind	6.46	1837	9.46	5.44	2183	11.24
120	Bare	No wind	7.40	1606	8.26	6.43	1846	9.50
140	Bare	No wind	7.72	1539	7.92	7.08	1678	8.64
167	Bare	No wind	8.15	1459	7.51	7.91	1501	7.73
212	Bare	No wind	8.85	1343	6.91	8.80	1351	6.95

Vertical Span Limited By Cross-Arm Strength

1 Tangent Structures
   Modified H-frame 26 ft. Arm
   Wood species is: Douglas Fir
   Fiber stress is: 8000 psi
   Safety factor = 4
   Number of arms used in calculations = 1
   X-arm cross section: Width ($5^{3}/_{8}''$), Depth ($7^{5}/_{8}''$)
   $^{1}/_{8}''$ will be subtracted from the above dimensions

$$\text{Max. bending moment} = \text{(Fiber stress w/S F) (width)} \times \text{(Depth * * 2)/(6)}$$

$$= 2000.0 \times 1 \times 5.250 \times 7.500 \times 7.500/6$$

$$= 98437.6 \text{ in.-lb}$$

$$\text{Vertical span} = \text{([Moment/Lever arm]} - \text{Ins weight) /} \text{Cable weight with Ice}$$

$$= ([98437.6 \ / \ 7.50] - 78.0) \ / \ 1.50120$$

$$= 822.3 \text{ feet (use maximum temperature curve)}$$

Span Limited by Conductor Separation

1 Tangent structure : Modified H-frame

$$\text{Horizontal separation} = \text{(0.025) (kV)} + \text{(F) Sq. root [60° Sag])} + \text{(.71) (ins length)}$$

$$12.50 = 0.025 \ (115.0 \text{ kV}) + 1.25 \ \text{(Sq. root) [60° Sag])} + 0.71 \ (3.93)$$

$$60° \text{ Final sag} = 29.90$$

$$\text{Maximum span} = \text{Sq. root } (380 \times 380 \times 29.90/5.38)$$

$$= 895.9 \text{ feet}$$

Conductor-Separation-At-Midspan

Structure-type modified H-frame

$$\text{OHGW} = {}^{3}/_{8} \text{ H.S. Steel}$$

$$\text{Conductor} = 477 \ 26/7 \text{ ACSR (Hawk)}$$

$$\text{Ruling span} = 380$$

1. Normal Transmission to OHGW @ 60° Final

   380 *Average span*

OHGW Sag	= (−) 2.76
Transmission sag (Bare)	= (+) 5.38
Separation at support	= (+) 10.00

Midspan separation	= 12.62 ft

   590 *Maximum span*

OHGW Sag	= (−) 6.65
Transmission sag (Bare)	= (+) 12.97
Separation at support	= (+) 10.00

Midspan separation	= 16.32 ft

2. Iced OHGW to Bare Transmission @ 32° Final

   380 *Average span*

OHGW Sag	= (−) 5.06
Transmission sag (Bare)	= (+) 4.36
Separation at support	= (+) 10.00

Midspan separation	= 9.30 ft

   590 *Maximum span*

OHGW Sag	= (−) 12.20
Transmission sag (Bare)	= (+) 10.51
Separation at support	= (+) 10.00

Midspan separation	= 8.31 ft

The Ratio of OHGW/Transmission conductor @ 60° final is:

   = 0.5130 for all span lengths

MW = Moment due to wind on wires

MP = Moment on pole

MR = Maximum allowable moment at ground line

SPAN = Sum of adjacent spans allowed

MW = Sum from top to ground of (Height above ground) × (Number of wires) × (Transverse wire force Wh)

MP   = ([Force of wind] × [Height of pole above ground squared] × [2 × Top dia. + Ground line dia.])/72

MR   = .000264 × Fiber stress = The cube of the pole cir. at ground line

SPAN = 2 × ([MR/Safety factor] − MP)/MW

Structure Type/Description: Modified H-Frame

Pole type is: Douglas Fir or S.Y.P.
Pole class is: 3
Pole safety factor used: 4
Pole buried standard depth + (0) feet
Heavy loading zone
Wire sizes are: ³/₈ H.S. Steel, Hawk
Bayonets are used

Pole Height	Dia. Gnd. Line	MW	MP	MR	SPAN
20	8.78	16.90	333.1	44,265.1	1,270
25	9.50	21.52	536.5	56,127.8	1,254
30	10.25	26.72	829.9	70,429.7	1,256
35	10.82	31.91	1,189.8	83,007.8	1,226
40	11.46	37.69	1,676.3	98,550.6	1,218
45	11.88	42.89	2,183.8	109,729.0	1,177
50	12.30	48.09	2,767.3	121,795.0	1,151
55	12.72	53.29	3,429.8	134,786.0	1,136
60	13.15	58.48	4,174.4	148,767.0	1,129
65	13.57	63.68	5,003.5	163,664.0	1,128
70	13.84	68.88	5,888.6	173,724.0	1,090
75	14.27	74.08	6,891.2	190,298.0	1,098
80	14.55	79.28	7,946.0	281,591.0	1,071
85	14.83	84.48	9,096.6	213,438.0	1,048
90	15.10	89.67	10,313.7	225,640.0	1,028
95	15.39	94.87	11,631.0	238,479.0	1,012
100	15.67	100.07	13,039.1	251,763.0	997
105	15.95	105.27	14,539.7	265,480.0	985
110	16.23	110.47	16,137.0	279,904.0	975
115	16.51	115.67	17,829.7	294,685.0	966
120	16.64	120.87	19,529.4	301,922.0	926
125	16.93	126.06	21,414.9	317,599.0	920

Insulator Swing Calculations

$$
\begin{aligned}
\text{MAX} &= \text{Maximum allowable swing angle} \\
\text{NOR} &= \text{Normal} \\
\text{MIN} &= \text{Minimum} \\
\text{NEG} &= \text{Negative} \\
\text{H} &= \text{Horizontal span (assumed values)} \\
\text{V} &= \text{Vertical span values (calculated)} \\
\text{T(167F)} &= \text{Tension, 167 Deg. Final, Bare cond.} = 1{,}459.0 \text{ lb} \\
\text{T(60F)} &= \text{Tension, 60 Deg. Final, Bare cond.} = 2{,}205.0 \text{ lb} \\
\text{T(}-20\text{I)} &= \text{Tension, } -20 \text{ Deg. Initial, Bare cond.} = 4{,}642.0 \text{ lb}
\end{aligned}
$$

WH (Wind force) = 9.0 lb/sq. ft = 0.64320 lb/ft

$$
\begin{aligned}
\text{WV} &= \text{Weight per foot of bare conductor} = 0.65680 \text{ lb s/ft}
\end{aligned}
$$

Dia. of conductor = 0.85760 in.

$$
\begin{aligned}
\text{WI} &= \text{Weight of insulator string} \\
\text{ANG} &= \text{Deflection angle on structure at PI location}
\end{aligned}
$$

Maximum Swing Formula:

$$
V = \frac{\dfrac{[2 \times T(-20I) \times SIN(ANG/2)] + (H \times WH)}{TAN(MAX)} - (WI/2)}{(WV)}
$$

Normal Swing Formula:

$$
V = \frac{\dfrac{[2 \times T(60) \times SIN(ANG/2)]}{TAN(NOR)} - (WI/2)}{(WV)}
$$

Minimum Swing Formula:

$$
V = \frac{\dfrac{[2 \times T(167F) \times SIN(ANG/2)] - (H \times WH)}{TAN(MIN)} - (WI/2)}{(WV)}
$$

Negative Swing Formula:

$$V = \cfrac{\cfrac{[-2 \times T(167F) \times SIN(ANG/2)] - (H \times WH)}{TAN(NEG)} - (WI/2)}{(WV)}$$

Structure Type	No. of Ins	MAX.	NOR.	MIN.	NEG.
Modified H-Frame	7	58.10	28.50	0.00	0.00

MAXIMUM FORMULA : (COLD CURVE) MAX = 58.1 Degrees WI = 78 lb for modified H-Frame

Normal formula : (60F) NOR = 28.5 Degrees

					Vertical Span					
		(			Horizontal Span					)
Degrees	NOR.	200	300	400	500	600	700	800	900	1000
0.00	−59	63	123	184	245	306	367	428	489	550
0.50	−5	101	162	223	284	345	406	467	528	589
1.00	49	139	200	261	322	383	444	505	566	627
1.50	102	178	239	300	361	422	482	543	604	665
2.00	156	216	277	338	399	460	521	582	643	704
2.50	210	254	315	376	437	498	559	620	681	742
3.00	264	293	354	415	476	537	598	659	720	780
3.50	318	331	392	453	514	575	636	697	758	819
4.00	372	370	431	492	552	613	674	735	796	857
4.50	426	408	469	530	591	652	713	774	835	896
5.00	480	446	507	568	629	690	751	812	873	934

69 kV TRANSMISSION LINE
TS–1 1/0 6/1 500 ft ruling span 1952

Station No. − 58207

Sum of spans = 57945

The cube of the spans = 8036988038.5

Longest span = 590

Shortest span = 201

Average span = 362.15625

The actual Ruling Span is = 372.425

Station No. − 60903.6015625

$$\text{Sum of spans} = 2697$$
$$\text{The cube of the spans} = 410138328$$
$$\text{Longest span} = 448$$
$$\text{Shortest span} = 339$$
$$\text{Average span} = 337.125$$
$$\text{The actual Ruling Span is} = 389.964$$

Station No. − 130768

$$\text{Sum of spans} = 70065$$
$$\text{The cube of the spans} = 9872838406$$
$$\text{Longest span} = 437$$
$$\text{Shortest span} = 260$$
$$\text{Average span} = 370.7142857142857$$
$$\text{The actual Ruling Span is} = 375.38$$

Station No. − 175272

$$\text{Sum of spans} = 44104$$
$$\text{The cube of the spans} = 6129378527$$
$$\text{Longest span} = 445$$
$$\text{Shortest span} = 206$$
$$\text{Average span} = 364.495867768595$$
$$\text{The actual Ruling Span is} = 372.794$$

Station No. − 182178

$$\text{Sum of spans} = 6806$$
$$\text{The cube of the spans} = 998855518$$
$$\text{Longest span} = 443$$
$$\text{Shortest span} = 305$$
$$\text{Average span} = 358.2105263157895$$
$$\text{The actual Ruling Span is} = 383.094$$

Appendix C

U.S.-Metric
Relationships

	U.S. To Metric		Metric To U.S.	
Length				
1 inch	=	25.4 mm	1 millimetre	= 0.03937 inch
1 inch	=	2.54 cm	1 centimeter	= 0.3937 inch
1 inch	=	0.0254 m	1 metre	= 39.37 inch
1 foot	=	0.3048 m	1 metre	= 3.2808 feet
1 yard	=	0.9144 m	1 metre	= 1.094 yard
1 mile	=	1.609 km	1 kilometre	= 0.6214 mile
Surface				
1 inch^2	=	645.2 mm^2	1 millimetre^2	= 0.00155 inch^2
1 inch^2	=	6.452 cm^2	1 centimetre^2	= 0.155 inch^2
1 foot^2	=	0.0929 m^2	1 metre^2	= 10.764 foot^2
1 yard^2	=	0.8361 m^2	1 metre^2	= 1.196 yard^2
1 acre	=	0.4047 hectare	1 hectare	= 2.471 acres
1 mile^2	=	258.99 hectare	1 hectare	= 0.00386 mi^2
1 mile^2	=	2.59 km^2	1 kilometre^2	= 0.3861 mile^2
Volume				
1 inch^3	=	16.39 cm^3	1 centimetre^3	= 0.061 inch^3
1 foot^3	=	0.0283 m^3	1 metre^3	= 35.314 foot^3
1 yard^3	=	0.7645 m^3	1 metre^3	= 1.308 yard^3
1 foot^3	=	28.32 litres	1 litre	= 0.0353 foot^3
1 inch^3	=	0.0164 litre	1 litre	= 61.023 inch^3
1 quart	=	0.9463 litre	1 litre	= 1.0567 quarts
1 gallon	=	3.7854 litres	1 litre	= 0.2642 gallons
1 gallon	=	0.0038 m^3	1 metre^3	= 264.17 gallons
Weight				
1 ounce	=	28.35 grams	1 gram	= 0.0353 ounce
1 pound	=	0.4536 kg	1 kilogram	= 2.2046 lb*
1 net ton	=	0.9072 T (metric)	1 Ton (metric)	= 1.1023 net tons**
*Avoirdupois				
**1 ton = 2000 lb				
Compound units				
1 lb/ft	=	1.4882 kg/m	1 kilogram/metre	= 0.6720 lb/ft
1 lb/in^2	=	0.0703 kg/m^2	1 kg/cm^2	= 14.223 lb/in^2
1 lb/ft^2	=	4.8825 kg/m^2	1 kg/m^2	= 0.2048 lb/ft^2
1 lb/ft^3	=	16.0192 kg/m^3	1 kg/m^3	= 0.0624 lb/ft^3
1 ft-lb	=	0.1383 kg-m	1kg-m	= 7.233 ft-lbs
1 hp	=	0.746 kW	1 kW	= 1.340 hp
1 ft-lb/in^2	=	0.0215 kg-m/cm^2	1 kg-cm/m^2	= 46.58 ft-lb/in^2
Temperature				
1 degree F	=	5/9 degree C	1 degree C	= 9/5 degree F
Temp °F	=	$\frac{9}{5}$ °C + 32	Temp °C	= $\frac{5}{9}$ (°F − 32)

Index